OXYGEN

By the same Authors

Carl Djerassi

Fiction
The Futurist and Other Stories
Cantor's Dilemma
The Bourbaki Gambit
Marx, Deceased
Menachem's Seed
NO

Poetry
The Clock Runs Backward

Drama
An Immaculate Misconception
Oxygen (with Roald Hoffmann)

Nonfiction
The Politics of Contraception
Steroids Made It Possible
The Pill, Pygmy Chimps, and Degas' Horse
From the Lab into the World: A Pill for People, Pets, and Bugs
This Man's Pill: Reflections on the 50th Birthday of the Pill

Scientific Monographs
Optical Rotatory Dispersion: Applications to Organic Chemistry
Steroid Reactions: An Outline for Organic Chemists (editor)
Interpretation of Mass Spectra of Organic Compounds (with H. Budzikiewicz and D. H. Williams)
Structure Elucidation of Natural Products by Mass Spectrometry (2 volumes with H. Budzikiewicz and D. H. Williams)
Mass Spectrometry of Organic Compounds (with H. Budzikiewicz and D. H. Williams)

Roald Hoffmann

Poetry
The Metamict State
Gaps and Verges
Memory Effects

Drama
Oxygen (with Carl Djerassi)

Nonfiction
Chemistry Imagined (with Vivian Torrence)
The Same and Not the Same
Old Wine, New Flasks: Reflections on Science and Jewish Tradition (with Shira Leibowitz Schmidt)

Scientific Monographs
The Conservation of Orbital Symmetry (with R. B. Woodward)
Solids and Surfaces: A Chemist's View of Bonding in Extended Structures

OXYGEN

A play in 2 acts

BY

CARL DJERASSI

AND

ROALD HOFFMANN

WILEY-VCH

Weinheim · New York · Chichester
Brisbane · Singapore · Toronto

Authors

Carl Djerassi
Department of Chemistry
Stanford University
Stanford, CA 94305–5080
e-mail: djerassi@stanford.edu
url: http://www.djerassi.com

Roald Hoffmann
Department of Chemistry
Cornell University
Ithaca, NY 14853–1301
e-mail: rh34@cornell.edu

Library of Congress Card No.:
applied for

British Library Cataloguing-in-Publication Data
A catalogue for this book is available from the British Library

Die Deutsche Bibliothek – CIP Cataloguing-in-Publication Data
A catalogue record for this publication is available from Deutsche Bibliothek

Printed in Federal Republic of Germany

Printed on acid-free paper

Composition Typomedia, Ostfildern
Cover Design Gunther Schulz
Printing and Bookbinding
Franz Spiegel Buch GmbH, Ulm
ISBN 3-527-30413-4

FOREWORD

What is discovery? Why is it so important to be first? These are the questions that trouble the people in this play. "Oxygen" alternates between 1777 and 2001–the Centenary of the Nobel Prize–when the Nobel Foundation decides to inaugurate a "retro-Nobel" Award for those great discoveries that preceded the establishment of the Nobel Prizes one hundred years before. The Foundation thinks this will be easy, that the Nobel Committee can reach back to a period when science was done for science's sake, when discovery was simple, pure, and unalloyed by controversy, priority claims, and hype . . .

The Chemistry Committee of the Royal Swedish Academy of Sciences decides to focus on the discovery of Oxygen, since that event launched the modern chemical revolution. But who should be so honored? Lavoisier is a natural choice, for if there ever was a marker for the beginning of modern chemistry, it was Lavoisier's understanding of the true nature of combustion, rusting, and animal respiration, and the central role of oxygen in each of these processes, formulated in the period 1770–1780. But what about Scheele? What about Priestley? Didn't they first discover oxygen?

Indeed, on an evening in October 1774, Antoine Lavoisier, the architect of the chemical revolution, learned that the Unitarian English minister, Joseph Priestley, had made a new gas. Within a week, a letter came to Lavoisier from the Swedish apothecary, Carl Wilhelm Scheele, instructing the French scientist how one might synthesize this key element in Lavoisier's developing theory, the lifegiver oxygen. Scheele's work was carried out years before, but remained unpublished until 1777.

Scheele and Priestley fit their discovery into an entirely wrong logical framework – the phlogiston theory – that Lavoisier is about to demolish. How does Lavoisier deal with the Priestley and Scheele discoveries? Does he give the discoverers their due credit? And what is discovery after all? Does it matter if you do not fully understand what you have found? Or if you do not let the world know?

In a fictional encounter, the play brings the three protagonists and their wives to 1777 Stockholm at the invitation of King Gustav III (of *Un ballo in maschera* fame). The question to be resolved: "Who discovered oxygen?" In the voices of the scientists' wives, in a sauna and elsewhere, we learn of their lives and those of their husbands. The actions of Mme. Lavoisier, a remarkable woman, are central to the play. In the Judgment of Stockholm, a scene featuring chemical demonstrations, the three discoverers of oxygen recreate their critical experiments. There is also a verse play within a play, on the Victory of Oxygen over Phlogiston. Such a play, now lost, was actually staged by the Lavoisiers for their friends and patrons.

Meanwhile, in the beginning of the 21st century, the Nobel Committee investigates and argues about the conflicting claims of the three men. Their discussions tell us much about whether science has changed in the last two centuries. The chair of the Nobel Committee is Astrid Rosenqvist, an outstanding Swedish theoretical chemist, while a young historian, Ulla Zorn, serves as a recorder for the committee's proceedings. But with time, her role changes.

The ethical issues around priority and discovery at the heart of this play are as timely today as they were in 1777. As are the ironies of revolutions: Lavoisier, the chemical revolutionary, is a political conservative, who loses his life in the Jacobin terror. Priestley, the political radical who is hounded out of England for his support of the French revolution, is a chemical conservative. And Scheele just wants to run his pharmacy in Köping, and do chemical experiments in his spare time. For a long time, he – the first man on earth to make oxygen in the laboratory – got least credit for it. Will that situation be repaired 230 years after his discovery?

PRODUCTION HISTORY

The World Premiere of "OXYGEN" was produced by the San Diego Repertory Theatre
(Sam Woodhouse, Artistic Director; Karen Wood, Managing Director)
April 2, 2001 in the Lyceum Theatre, San Diego, CA.

Director	Brian Bevell
Set Designer	David Weiner
Costume Designer	Melanie Watnick
Scenic designer	David Cuthber
Sound designer	Todd Reischman

A Workshop Production of 10 performances was staged by the Eureka Theatre
(Andrea Gordon, Co-Artistic Director)
May 2000, San Francisco, CA

Director	Andrea Gordon
Set Designer	Kelly Hatch
Costume Designer	Jean Frederickson
Lighting Designer	Tiffani Snow
Composer	Derrick Okubo
Stage Manager	Heather Beckett*

Marie Anne Lavoisier	Lilly Akseth
Joseph Priestley/Ulf Svanholm	David E. Kazanjian*
Mary Priestley/Astrid Rosenqvist	Barbara Early*
Carl Wilhelm Scheele/Sune Kallstenius	Allen McElvey*
Sara Margaretha Pohl/Ulla Zorn	Amy Mordecai
Antoine Lavoisier/Bengt Hjalmarsson	Simon Vance*
Voice of Court Herald	Dawson Moore

* Members, Actors Equity Association

CAST OF CHARACTERS

Stockholm, 1777

ANTOINE LAURENT LAVOISIER
34 years old. *(French chemist, tax collector, economist, and public servant; discovered oxygen).*

MARIE ANNE PIERRETTE PAULZE LAVOISIER
19 years old. *(Wife of the above).*

JOSEPH PRIESTLEY
44 years old. *(English minister and chemist; discovered oxygen).*

MARY PRIESTLEY
35 years old. *(Wife of the above).*

CARL WILHELM SCHEELE
35 years old. *(Swedish apothecary; discovered oxygen).*

SARA MARGARETHA POHL (FRU POHL)
26 years old. *(Became* MRS. SCHEELE *three days prior to Carl Wilhelm's death).*

COURT HERALD
(off-stage male voice).

Stockholm, 2001

Prof. BENGT HJALMARSSON
member of the Chemistry Nobel Prize Committee of the Royal Swedish Academy of Sciences. *(Same actor as* ANTOINE LAVOISIER).

Prof. SUNE KALLSTENIUS
member of the Chemistry Nobel Prize Committee of the Royal Swedish Academy of Sciences. *(Same actor as* CARL WILHELM SCHEELE).

Prof. ASTRID ROSENQVIST
chair of the Chemistry Nobel Prize Committee of the Royal Swedish Academy of Sciences. *(Same actress as* MRS. PRIESTLEY).

Prof. ULF SVANHOLM
member of the Chemistry Nobel Prize Committee of the Royal Swedish Academy of Sciences. *(Same actor as* JOSEPH PRIESTLEY).

ULLA ZORN
a graduate student in the History of Science and amanuensis to the Chemistry Nobel Prize Committee. *(Same actress as* FRU POHL).

Technical Details

The staging can be sparse *(sauna bench; conference table; laboratory demonstration table)*. All audiovisuals, available from the authors, are to be projected on a large screen, preferably by rearward projection. To provide for rapid alternating costume changes between 1777 and 2001, the 1777 dress code should be distinctive yet simple (e.g. use of wigs, long coats with easily attachable ruffled collars for men; distinctive (buckled) shoes; wigs, mobcaps, scarves, long dresses for women, etc).

SCENE 1

(Sauna in Stockholm, Sweden, 1777). The three women sit on a sauna bench, their bodies covered to various extents by bathing towels or appropriate sheets – Mrs. Priestley most decorously and Mme. Lavoisier most daringly. Each is wearing a different, typically 18th century, mobcap to cover her hair or wig).

MME. LAVOISIER

(Dreamily)

I have never been beaten before . . . not like that. Can we do it again?

MRS. PRIESTLEY

Madame! In England the birch is used for chastisement.

FRU POHL

In Sweden, we consider it healthy. It brings the blood to the surface. So much better than leeches.

MRS. PRIESTLEY

(As towel slips off her shoulder, she quickly pulls it up)

The immodesty of the sauna disquiets me.

MME. LAVOISIER

(Deliberately lowers her towel while addressing Mrs. Priestley)

Mrs. Priestley . . . we are just women. *(Aside)* . . . Now, were there men here . . .

MRS. PRIESTLEY

Oh, you are young, Madame!

MME. LAVOISIER

Nineteen!

FRU POHL

I was twenty before I married.

MRS. PRIESTLEY

So was I.

(Turns to Fru Pohl)

How many children do you have?

FRU POHL

A young son. And you?

MRS. PRIESTLEY

Three sons and a daughter.

(Turns to Mme. Lavoisier)

And you, Madame Lavoisier?

MME. LAVOISIER

None.

MRS. PRIESTLEY

Ah! I presume you married only recently?

MME. LAVOISIER

Six years ago.

FRU POHL

And no children?

MRS. PRIESTLEY

My first child was born when we were married but ten months –

MME. LAVOISIER

As we say in France, *chacun à son goût.*

MRS. PRIESTLEY

So you think it was a matter of taste? I took it as an obligation when I married.

(A touch of sarcasm)

But then, of course, I was twenty ...

MME. LAVOISIER

Perhaps women mature faster in France ... especially those brought up in convent schools.

MRS. PRIESTLEY

A convent?

MME. LAVOISIER

Not to become a nun. And when my mother died, I left the convent to serve as my father's hostess. I was twelve. *(Pause)*. I even studied chemistry... "Butter of arsenic"... "Sugar of lead"... "Flowers of zinc." What wonderful words, I thought: First chemistry in the kitchen ... then chemistry in the garden ...

MRS. PRIESTLEY

A child of twelve would find it charming.

MME. LAVOISIER

At thirteen, I escaped the attentions of a Count – much older than my father – by marrying Monsieur Lavoisier. *(Proudly)*. He is active in the tax collection agency for the crown. He heads the Discount Bank –

MRS. PRIESTLEY

A tax collector? A banker?

MME. LAVOISIER

(Amused)

And a lawyer at twenty-one!

FRU POHL

Yet your husband was invited to Sweden because of his chemical discoveries?

MME. LAVOISIER

So was Mrs. Priestley's husband.

(Disingenuously to Mrs. Priestley).

He is a priest, is he not?

MRS. PRIESTLEY

A minister. Whom people call "Dr." Priestley.

(Suddenly agitated)

When you marry a man of God, you know you will find riches greater than money. But our Unitarian ideas are opposed by the Church of England. We cannot hold government office, we cannot go to Oxford or Cambridge. *(Catches herself)*. I beg your pardon ... I was carried away.

MME. LAVOISIER

When I spoke of the chemistry I learned in the convent ... my husband told me something very useful. "The product of science is knowledge ... but the product of scientists is reputation." *(Pause)*. Reputation is important to him ... and when I married him, it also became important to me. *(Pause)*. Especially when he asked me to assist him in his endeavors.

FRU POHL

He asked you that ... at age thirteen?

MME. LAVOISIER

Bien sûr ... There was chemistry to study. Art too. I took lessons with Jacques-Louis David ... all to help my husband.

(She muses)

Each day in the laboratory, I made a list of what experiments were to be done. Antoine called out the numbers, I wrote them down. I drew the plates for his books ... I etched them ... I corrected them.

MRS. PRIESTLEY

(Suddenly compassionate)

Is that why you have no children?

MME. LAVOISIER

(Ignores the comment)

There was Latin to learn, and English too. It is I, Mrs. Priestley, who translated Dr. Priestley's *Experiments on Different Kinds of Airs* ... and his writings on phlogiston –

MRS. PRIESTLEY

(Quickly interrupts)

The principle of fire ... an explanation for all chemistry.

MME. LAVOISIER

His explanation.

MRS. PRIESTLEY

What do you mean?

MME. LAVOISIER

We are not convinced –

MRS. PRIESTLEY

We?

MME. LAVOISIER

My husband is not convinced ... and therefore, I am not convinced.

FRU POHL

Herr Scheele is convinced. He says so in his book . . .

MME. LAVOISIER

(Very curious)

What book?

FRU POHL

The only book he has written. On the chemistry of air and fire.

MME. LAVOISIER

My husband never mentioned it.

FRU POHL

It will come out soon . . . perhaps while you are still in Stockholm.

MME. LAVOISIER

(Relieved)

So this is your husband's newest work?

FRU POHL

Apothecary Scheele is not my husband . . .

MRS. PRIESTLEY

I thought Pohl was your father's name . . .

FRU POHL

Herr Pohl was an apothecary. And the father of my son. But he's dead.

MRS. PRIESTLEY

(Unable to restrain her curiosity)

And Monsieur Scheele? Perhaps he is a relative?

FRU POHL

He took over my husband's pharmacy . . . in Köping . . . some thirty leagues west of Stockholm. Where I keep house for him.

MME. LAVOISIER

You do assist Monsieur Scheele?

FRU POHL

Not in the laboratory.

MME. LAVOISIER

Yet you know of his new book?

FRU POHL

When Carl Wilhelm *(catches herself)* . . . I mean Apothecary Scheele . . . arrived in our little town two years ago, he related his work on airs to my father and me. He was so excited about it.

MME. LAVOISIER

(Taken aback)

And when was that work done?

FRU POHL

Some years earlier, I'm sure. The book tells it all . . .

MME. LAVOISIER

Are its contents known to anyone?

FRU POHL

Your husband. *(Pause)*. Did Herr Scheele not send three years ago a letter to Paris describing his experiment with Fire Air?

MME. LAVOISIER

I know of no correspondence between them.

FRU POHL

I must confess he wondered why your husband never thanked him . . .

MME. LAVOISIER

(Agitated)

He had nothing to thank him for!

MRS. PRIESTLEY

(Trying to calm things down)

Ladies . . . perhaps we should cool off.

FRU POHL

You are right. Come.

(Rises, stretches out hand to MME. Lavoisier while reaching for birch branch still in Mrs. Priestley's hand)

You have sweated enough. Mrs. Priestley's birch is waiting.

END OF SCENE I

INTERMEZZO 1

Immediately following Scene 1
(Left downstage, very dark, spotlight solely on face)

MME LAVOISIER
(Mimics Fru Pohl's voice and intonation)
"And no children?"
(Resumes normal voice and accent)
What gives Fru Pohl the right to ask?... Not even married to Apothecary Scheele!
(Pause)
I helped Antoine in the laboratory... as in the salon. But when he reasoned out how we breathe... how sulfur burns... how to make better gunpowder... he spoke to men: to Monsieur Monge... to Monsieur Laplace... to Monsieur Turgot. *(Pause)*. But not to me.
(Pause)
Yet I helped Antoine in ways he doesn't know about... and never will.
(Pause)
But I must be careful with Mme. Priestley... and now, I see, also with Mme. Pohl. We did not come to Stockholm to make mistakes. So... we talk women's talk. About our husbands, of course. How good they are. How we help them.
(Pause)
Wearing the woman's mask... her husband's face on it... smiling politely.
(Pause)
But will the men go on smiling when their discoveries are disputed?
(Pause)
Will we?
(Collecting herself from a daydream, more forcefully).
She knows of the letter, our Mme. Pohl. *(Pause)*. I am afraid.

BLACKOUT

SCENE 2

(Conference room at Royal Swedish Academy of Sciences, Stockholm, summer 2001. Lights focus on two members of Nobel Committee for Chemistry, Professors BENGT HJALMARSSON *and* SUNE KALLSTENIUS, *who are huddling left downstage in almost whispered private conversation. Later on, the third member,* ULF SVANHOLM, joins them.)

SUNE KALLSTENIUS

A retro-Nobel? There must be better ways of celebrating the centenary of the Nobel Prizes than establishing a new one for work done before 1901 . . .

BENGT HJALMARSSON

With no one alive to receive it. *(Pause)*. But now?

SUNE KALLSTENIUS

I rather fancy recognizing dead people – it's different.

BENGT HJALMARSSON

It's still a lot of work.

SUNE KALLSTENIUS

You always complain about the time spent on Nobel Committee business.

BENGT HJALMARSSON

All I seem to be doing is reading other people's papers.

SUNE KALLSTENIUS

How else can we come up with a list of candidates?

BENGT HJALMARSSON

What about my important work?

SUNE KALLSTENIUS

Most Swedes would be proud to pay that price!

BENGT HJALMARSSON

I'm tired of paying it! No wonder Swedish chemists don't win the real Prize!

SUNE KALLSTENIUS

What about Tiselius?

BENGT HJALMARSSON

(Dismissive)

50 years ago!

SUNE KALLSTENIUS

What about Bergström? What about Samuelsson?

BENGT HJALMARSSON

That was in Medicine. And they shared it.

SUNE KALLSTENIUS

So resign.

BENGT HJALMARSSON

(Grins)

From the committee? I like the power . . . and the gossip.

SUNE KALLSTENIUS

Now you have double power: Picking regular Nobel prizewinners as well as retro-Nobelists. First the living . . . now the dead.

BENGT HJALMARSSON

The dead don't repay favors.

SUNE KALLSTENIUS

Do you want people to hear that?

BENGT HJALMARSSON

I'm just being honest.

SUNE KALLSTENIUS

Honesty has its place . . . but this isn't it!

(Ulf Svanholm, entering, overhears this)

ULF SVANHOLM

I'm surprised to hear that from you . . . of all people.

SUNE KALLSTENIUS

(Sharply)

You would say that!

BENGT HJALMARSSON

(pensive)

Astrid as chairman of a Nobel committee–

ULF SVANHOLM

She prefers to be called "chair."

BENGT HJALMARSSON

We've never before had a woman . . .

SCENE 2

ULF SVANHOLM

She deserves it; a damned good theoretician . . .

SUNE KALLSTENIUS

In my experience, theoreticians make lousy chairmen.

BENGT HJALMARSSON

I wouldn't generalize when it comes to Astrid. Besides, she always gets her way.

ULF SVANHOLM

How do you know?

BENGT HJALMARSSON

Take my word for it. I know.

ULF SVANHOLM

Oh, I forgot! The two of you had something going . . .

BENGT HJALMARSSON

That was nearly eighteen ago. *(Pause)*. There she comes . . . with that mysterious Ulla Zorn.

CROSSFADE *to right downstage. Professor* ASTRID ROSENQVIST, *chair of the committee, and* ULLA ZORN, *approach in almost whispered conversation.*

ULLA ZORN

You haven't told them about me, have you?

ASTRID ROSENQVIST

Not yet, Ulla.

ULLA ZORN

They must be wondering–

ASTRID ROSENQVIST

I'm sure they are. Nobel Committee secretaries are usually older.

ULLA ZORN

Aren't they expecting a chemist for secretary?

ASTRID ROSENQVIST

That's why you are called an amanuensis.

ULLA ZORN

Why not tell them what I do? It's no secret –

ASTRID ROSENQVIST

All in good time . . . trust me. *(Pause).* Look, the men are already here.

(Looks at watch, approaches men, addresses Bengt Hjalmarsson)

You're early–

BENGT HJALMARSSON

No, we're punctual . . . like all Swedes. Your watch needs fixing.

ASTRID ROSENQVIST

(Smiling, but sharp edge)

You haven't changed, Bengt. Always the last word.

(To rest of group)

Let's sit down and get to work.

(Committee members move to conference table, with ULLA ZORN, *laptop computer in front of her, sitting somewhat separately on one side. Large name signs in front of each committee member may be helpful for audience).*

SUNE KALLSTENIUS

(Addresses Astrid Rosenqvist)

A procedural question: why are there only four of us? We've never had fewer than five members. You have no deadlocks with an odd-numbered committee.

ULF SVANHOLM

Leave it to Sune . . . always complaining.

ASTRID ROSENQVIST

There's nothing magic about the number five. There's no precedent for what we're asked to do . . .

BENGT HJALMARSSON

You can say that again: restricting our choices to the 19th century or earlier!

SUNE KALLSTENIUS

At least we have fewer Americans. In fact, only one: Willard Gibbs. What's chemistry without thermodynamics . . . without the Phase Rule?

ULF SVANHOLM

For this . . . the first retro-Nobel? And again an American? *(Pause)*. The choice is obvious.
(Slow and forceful).
Dimitri . . . Ivanovitch . . . Mendeleyev. Can you imagine chemistry without the Periodic Table? It's our Rosetta Stone.

BENGT HJALMARSSON

What about Louis Pasteur?
(Speaks slowly and pompously)
"The Prizes should be distributed to those who have conferred the greatest benefit on Mankind."
(Reverts to ordinary tone)
That's what it says in Alfred Nobel's Will. *(Pause)*. If you stop people on the street with the question, "Who has conferred the greatest benefit on Mankind? Gibbs? Mendeleyev? Or Pasteur?" They'll say, "Gibbs? Never heard of him! Mendeleyev? Spell it!" Everyone knows Pasteur.

ULF SVANHOLM

But we aren't people on the street!
(Suddenly notices Ulla Zorn furiously typing on her keyboard)
Wait a moment!
(Points to Ulla Zorn)
Is this part of the formal meeting?

ASTRID ROSENQVIST

Everything is being put down for the record.

ULF SVANHOLM

But why?

ASTRID ROSENQVIST

With our regular Nobel, each year, we solicit thousands of nominations from all over the world . . .

BENGT HJALMARSSON

Thank God most of them are too lazy to respond.

ULF SVANHOLM

But why the computer?

ASTRID ROSENQVIST

Because we aren't just preparing the usual recommendation to the Academy on who should get the Prize . . . we're also generating the pool of candidates. We need a record . . . to show that it was all above board.

BENGT HJALMARSSON

I'm still amazed that we were asked to do both.

ASTRID ROSENQVIST

The whole retro-Nobel announcement is supposed to come as a surprise. How can you do that by broadcasting that we want a list of candidates?

(Taps on table)

We have Gibbs, Mendeleyev, Pasteur . . . *(Pause)*. What other names would you like to throw into the pot?

ULF SVANHOLM

Why not a Swede for the first one? When it came to the regular Nobel Prizes, the Academy waited until 1903 before giving it to Arrhenius.

ASTRID ROSENQVIST

He can't just be Swedish! He also has to deserve it.

BENGT HJALMARSSON

How about Carl Wilhelm Scheele . . . for the discovery of oxygen–

ULF SVANHOLM

Start with the 18th century?

SUNE KALLSTENIUS

(Cynical, pointing to Ulf)

Probably he wants to give it to Paracelsus!

ASTRID ROSENQVIST

No retro-Nobel for alchemists.

BENGT HJALMARSSON

Focusing on the 18th century may not be a bad idea. People published less . . . so we have less to read.

ULF SVANHOLM

But if we select Scheele, what about Lavoisier?

SUNE KALLSTENIUS

Or Joseph Priestley?

BENGT HJALMARSSON

Right back to the usual Nobel quandary! Too many candidates.

ULF SVANHOLM

How about John Dalton, the father of the atomic theory?

SUNE KALLSTENIUS

That's not logical. Oxygen had to be discovered first . . . and its role in chemistry understood! Maybe for the second or third retro-Nobel . . .

ASTRID ROSENQVIST

Sune has a point; the Chemical Revolution came from oxygen. The element ought to come first.

ULF SVANHOLM

Even if a Frenchman or an Englishman gets the credit?

SUNE KALLSTENIUS

Gets the credit? Surely you mean shares it!

ASTRID ROSENQVIST

That's up to our committee to determine . . .

BENGT HJALMARSSON

And since there are no living contemporaries, we don't have to solicit the opinions of outside experts.

ULF SVANHOLM

We may have to turn to historians.

(Ulla Zorn looks up).

I'm joking.

ASTRID ROSENQVIST

What's wrong with historians?

SUNE KALLSTENIUS

It's a thing scientists do when they can't do science anymore.

ASTRID ROSENQVIST

But professional historians?

BENGT HJALMARSSON

What would they know about science? *(Pause)*. You might as well search the web!

ASTRID ROSENQVIST

(Looks at Ulla Zorn, but decides not to pursue her defense of historians)

I wonder whether Scheele, Lavoisier, and Priestley ever met in one place.

ULLA ZORN

Very unlikely.

BENGT HJALMARSSON

What makes you say that?

ULLA ZORN

Absence of any evidence.

BENGT HJALMARSSON

But how would you know–

ASTRID ROSENQVIST

(Quickly cuts off further question)

Just think of the royal competitions of that time . . . Maybe they met at some 18th century forerunner to our modern Nobel Prizes. Why not in Stockholm? We had a king then, Gustav the Third, who was wild about science and the arts.

ULF SVANHOLM

(Bantering)

Keep on dreaming! And how would they've talked to each other?

ASTRID ROSENQVIST

(Returns banter)

Who worries about language in dreams?

SUNE KALLSTENIUS

Dr. Sigmund Freud.

ULF SVANHOLM

Maybe that's why he never got a Nobel Prize.

SUNE KALLSTENIUS

Ulf is always worried about missing out on prizes.

ASTRID ROSENQVIST

(Dismissive)

Sune, Ulf! It's time to bury the hatchet. *(Pause)*. But were they as ambitious as their modern successors? I wonder who could have told us?

ULF SVANHOLM

The obvious witnesses: other scientists of that time.

ULLA ZORN

Or their wives.

SUNE KALLSTENIUS

What did you say?

ULLA ZORN

Wives. *(Pause)*. Most men around that time had wives. Why not look for what they had to say?

END OF SCENE 2

INTERMEZZO 2

Immediately following Scene 2

(Stockholm, 1777, same day as Scene 1, a few hours later). Bare room, into which the three couples enter in turn, from left upstage, right downstage, and right upstage

Spotlight on MME. LAVOISIER *and* LAVOISIER *They whisper.*

MME. LAVOISIER
Beware!

LAVOISIER
Of what?

MME. LAVOISIER
A challenge.

LAVOISIER
An experiment?

MME. LAVOISIER
A book . . .

LAVOISIER
From Priestley?

MME. LAVOISIER
No, Scheele.

LAVOISIER
Scheele?

MME. LAVOISIER
Indeed.

LAVOISIER
He's a good chemist.

MME. LAVOISIER
And careful.

LAVOISIER
I trust him.

Spotlight focuses on MRS. PRIESTLEY *and* PRIESTLEY. *They whisper.*

MRS. PRIESTLEY

Take heed!

PRIESTLEY

Of what?

MRS. PRIESTLEY

An experiment.

PRIESTLEY

Mine's ready!

MRS. PRIESTLEY

It may have been done.

PRIESTLEY

By whom?

MRS. PRIESTLEY

Scheele.

PRIESTLEY

What can he have?

MRS. PRIESTLEY

Something from the past.

PRIESTLEY

He needs something new.

MRS. PRIESTLEY

He questions . . .

PRIESTLEY

I trust him.

Spotlight on FRU POHL *and* SCHEELE. *They whisper.*

FRU POHL

I told her.

SCHEELE

And?

FRU POHL

She denied it.

SCHEELE

He withheld it from her.

FRU POHL

I doubt it.

SCHEELE
Why?

FRU POHL
She keeps his correspondence.

SCHEELE
Ha!

FRU POHL
But she was most curious.

SCHEELE
And?

FRU POHL
She will tell her husband.

SCHEELE
I do not trust him.

BLACKOUT

SCENE 3

(Conference room at Royal Swedish Academy of Sciences, Stockholm, a few minutes after Scene 2. Lights focus on BENGT HJALMARSSON *and* ULF SVANHOLM, *who are huddling in almost whispered private conversation)*

BENGT HJALMARSSON

"Bury the hatchet." What was Astrid talking about?

ULF SVANHOLM

You don't know? Of course, Sune will deny it.

BENGT HJALMARSSON

(Impatiently)

Deny what?

ULF SVANHOLM

You remember the Stanford group's paper on new catalysts for oxygenated polymers?

BENGT HJALMARSSON

(Dismissive)

Didn't you have some similar catalysts up your sleeve?

ULF SVANHOLM

Identical. Except that the American paper came out several months earlier . . . and now they won the Gibbs Medal for that work . . . thanks to *(heavy sarcasm)* our distinguished colleague, Professor Kallstenius! I bet that's why he proposed Willard Gibbs for the retro-Nobel . . . just to rub it in.

BENGT HJALMARSSON

I don't get it.

ULF SVANHOLM

When I wrote up our work and sent it to the journal, Sune got it for review.

BENGT HJALMARSSON

So?

ULF SVANHOLM

He sat on it for two months before refereeing it.

BENGT HJALMARSSON

(Dismissive)

That's par for the course. Do you know how many articles I get to review?

ULF SVANHOLM

I wasted another half year getting some damned spectra he wanted. Meanwhile he told his Stanford pals in California all about it.

BENGT HJALMARSSON

(Turns serious)

Are you sure?

ULF SVANHOLM

Who else could have told them? He knows them . . . all too well!

BENGT HJALMARSSON

In research . . . simultaneous discovery occurs all the time.

ULF SVANHOLM

Stop preaching to me!

BENGT HJALMARSSON

Ulf, calm down! Why not assume they found it by themselves?

ULF SVANHOLM

Nonsense!

BENGT HJALMARSSON

You're obsessed by this. Let go.

ULF SVANHOLM

Obsessed? We're always in a race where being first counts for everything. If you're second, you might as well be last. There's only a Gold Medal–in this case the Gibbs Medal–but no silver or bronze.

BENGT HJALMARSSON

I wouldn't blame Sune. He's too honest . . . you just have to look at his face.

ULF SVANHOLM

I think you're on his side. We all wear masks.

BENGT HJALMARSSON

Which one is yours?

ULF SVANHOLM

Guess.

BLACKOUT

SCENE 3

(Stockholm, 1777, same day as Intermezzo 2, a few hours later).

SCHEELE

How gracious of you to travel so far, Monsieur Lavoisier. I've never left Sweden.

LAVOISIER

The invitation came from His Majesty. But –

SCHEELE

But, Monsieur?

LAVOISIER

His Majesty's curiosity on matters scientific is known to all of us . . .

SCHEELE

Indeed it is.

LAVOISIER

But does it encompass pneumatic chemistry?

SCHEELE

Perhaps.

LAVOISIER

(Sarcastic)

And includes a personal desire to have us verify in public, as the invitation states . . . "each savant's claims to Fire Air"?

SCHEELE

Perhaps it does.

LAVOISIER

One does not refuse a king. But –

SCHEELE

But, Monsieur?

LAVOISIER

Who is behind this? Who has the King's ear?

SCHEELE

Torbern Bergman. *Primus inter pares* among all Swedish scientists . . . as well as–

LAVOISIER

. . . your strongest patron.

SCHEELE

Surely not a fault?

LAVOISIER

We all have our patrons . . . and *(pretends to cross himself)* daily pray to God for their long life and continued support.

SCHEELE

What is your question then?

LAVOISIER

Bergman's genius classified all chemical matter into inorganic and organic . . .

SCHEELE

Only one of his many strokes of genius.

LAVOISIER

Professor Bergman has never concerned himself with airs. Why has he then arranged our meeting? To raise the Swedish flag above all others?

SCHEELE

Because he desires to know whom God's grace favored first among us three–

LAVOISIER

(Ironic)

While you do not?

SCHEELE

I already know. But–

LAVOISIER

But, Monsieur?

SCHEELE

But do you? *(Pause)*

(Priestley enters).

Or Dr. Priestley?

LAVOISIER

Ah, Monsieur. You arrived just in time.

(Addresses Priestley)

The royal invitation, you may recall, demands from each of us an actual experiment ...

PRIESTLEY

Indeed it does.

SCHEELE

Which, His Majesty suggests, will be executed by another.

PRIESTLEY

But why?

SCHEELE

To confirm each person's claim.

PRIESTLEY

Claim? Can what is fact be claimed?

SCHEELE

Once reproduced by another, claims become facts.

PRIESTLEY

So they do. But does the King, or *(Pause)* you doubt my experiments?

SCHEELE

No, my dear Doctor. But the world needs proof.

PRIESTLEY

Proof it shall have. Until tomorrow then!

LAVOISIER

(Stops him)

Un moment! Madame Lavoisier and I, desiring to divert you and your ladies ... and of course His Majesty ... have devised an entertainment for your pleasure ... *(Pause)* and perhaps enlightenment ... a play that we have written and have performed ...

(Pause) . . . but once. Would you permit us to present this evening a masque on phlogiston and his enemy?

PRIESTLEY

Ah, what strange ways of presenting scientific arguments you have in France!

LAVOISIER

But His Majesty, Gustavus the Third, loves masques!

SCHEELE

Perhaps too much . . . some say.

END OF SCENE 3

SCENE 4

(Stockholm, 2001; Royal Swedish Academy of Sciences, one week later. Committee members grouped around conference table, while Ulla Zorn sits with her computer on a separate small table).

ASTRID ROSENQVIST

First to the discovery: No one will question that oxygen confers great benefit on mankind, right?

BENGT HJALMARSSON

Oxygen was good for people before it was "discovered!"

ULF SVANHOLM

But there are plenty of benefits that require for oxygen to be isolated. What about the emphysema victim in an oxygen tent . . . the Everest climber with his oxygen bottles . . . the astronaut in the space suit?

SUNE KALLSTENIUS

We didn't pick oxygen for its value to mountain climbers or astronauts or sick people.

ULF SVANHOLM

There you go with your usual spiel . . . the academic's ivory tower disdain for the useful . . .

ASTRID ROSENQVIST

Let's compromise. Who'd like to come up with some simple phrases to explain to Ulf's public that without the discovery of oxygen there would've been no Chemical Revolution . . . no chemistry as we now know it?

BENGT HJALMARSSON

I'll give it a try. Prior to Lavoisier –

SUNE KALLSTENIUS

You mean prior to the discovery of oxygen –

BENGT HJALMARSSON

To me they are the same.

SUNE KALLSTENIUS

To me they are not.

BENGT HJALMARSSON

Never mind ... Before the Chemical Revolution, people were convinced that when things burned, something was released ... called phlogiston ...

(Turns to ZORN)

Do you want me to spell that?

ULLA ZORN

(Quick and dismissive, without looking up while typing quickly)

P ... H ... L ... O ... G ... I ... S ... T ... O ... N.

ASTRID ROSENQVIST

Hold it, Bengt! The public at large ... and these days, even many chemists ... won't have the slightest idea what phlogiston means. They can't even pronounce it. Please ... make it clear ... and make it short.

BENGT HJALMARSSON

"Phlogiston: The essence of fire." How's that for a pithy definition?

ASTRID ROSENQVIST

That's too pithy.

BENGT HJALMARSSON

You certainly are difficult to satisfy. But why even bother with a discarded theory?

ASTRID ROSENQVIST

Because Priestley and Scheele and most other 18th century chemists weren't fools. They believed in phlogiston till they died.

SUNE KALLSTENIUS

And it made sense ... in its own way. They thought when anything burns, something ... specifically that wondrous phlogiston ... leaves that burning object and goes out into the air.

ASTRID ROSENQVIST

For all of them, phlogiston represented the "Grand Unified Theory" of the chemistry of their time.

BENGT HJALMARSSON

(Sarcastic)

Oh, sure . . . it could account for anything. That supposedly commonsense theory was rudely punctured . . . by Lavoisier's revolutionary insight . . . that during the process of burning . . . something is taken from the air. And that "something" is oxygen!!

ULF SVANHOLM

Why not just say, the language of chemistry was a holy mess and the grammar all wrong? Let's get to the business of picking the winner. Prizes are given to people, not to discoveries.

ASTRID ROSENQVIST

Prizes go to people, sure. But they need to have discovered something, understood it.

(Pauses).

I now propose that each of you take the primary responsibility for digging up the evidence for the claims of one of the candidates. Who is fluent in French?

BENGT HJALMARSSON

Il n'y a pas de doute que c'est moi! I didn't spend two years as a postdoc at the Pasteur Institute speaking Swedish.

ASTRID ROSENQVIST

(Ignores comment)

Who else is fluent in French?

SUNE KALLSTENIUS

Try me in Greek or Latin. Or German . . .

ASTRID ROSENQVIST

(Addresses Svanholm)

And you?

ULF SVANHOLM

(Dismissive)

Comme ci, comme ça . . . usual high school French.

SUNE KALLSTENIUS

That's obvious.

ASTRID ROSENQVIST

The Lavoisier archives are mostly in France and, of course, written in French. Lavoisier is your man, Bengt.

(Turns to Kallstenius)

You know Scheele wrote mostly in German ... and some peculiar Latin? I propose you take Scheele ...

(Turns to Svanholm)

which leaves you with Priestley. OK?

ULF SVANHOLM

Are you offering me a choice?

ASTRID ROSENQVIST

I'm offering you a candidate. But if you're unhappy, you and Sune could collaborate on both men.

ULF SVANHOLM

Thanks! I'll take Priestley.

ASTRID ROSENQVIST

Of course, you could always have a duel.

SUNE KALLSTENIUS

Only if I can choose the weapon.

BENGT HJALMARSSON

Enough of that.

(Looks at watch and starts to get up)

Is that it for today?

ASTRID ROSENQVIST

There's one issue that absolutely requires digging into the original literature.

SUNE KALLSTENIUS

And what's that?

ASTRID ROSENQVIST

I'm referring to Scheele's letter to Lavoisier ... in which he outlined his own experiments with oxygen, which he called *Feuerluft* ... Did Lavoisier get that letter and if so, when?

ULF SVANHOLM

In other words, we're right back at our usual preoccupation with priority . . . the Nobel Syndrome: who did what first?

ASTRID ROSENQVIST

And did the one who did it first really know what he'd done?

ULF SVANHOLM

Why should that matter?

ASTRID ROSENQVIST

I'm a theoretician. For me it's necessary to understand what one finds. Maybe for you it matters less. *(Pause)*. You're an experimentalist . . . you actually get your hands dirty–

ULF SVANHOLM

Now it's my students' hands.

BENGT HJALMARSSON

So it's dirt we're looking for?

ULF SVANHOLM

I just wonder which kind we'll find . . . dirt from honest labor or the other sort?

BENGT HJALMARSSON

And where do we look?

ULLA ZORN

(Looks up from her PC)

The wives. *(Pause)*. That's where I would look.

ULF SVANHOLM

(Confused)

The wives?

ULLA ZORN

Aren't they usually expected to clean up the dirt?

END OF SCENE 4

INTERMEZZO 3

Immediately following Scene 4
(left downstage)

ULF SVANHOLM
Now what do you think of her?

BENGT HJALMARSSON
Astrid?

ULF SVANHOLM
No, Ulla Zorn.

BENGT HJALMARSSON
Deep water . . . and not still.

ULF SVANHOLM
Astrid called her an amanuensis.

BENGT HJALMARSSON
(Dismissive)
She was showing off. It's just a fancier word for secretary.

ULF SVANHOLM
Except for mentioning the wives . . . she's hardly said anything.

BENGT HJALMARSSON
That's why I'm suspicious.

ULF SVANHOLM
Of Zorn?

BENGT HJALMARSSON
Of Astrid. Springing Zorn on us is part of her private agenda. I can smell it!

ULF SVANHOLM
You keep talking about Astrid . . . I want to know what you think about the retro-Nobel.

BENGT HJALMARSSON
It's too early to tell. And you?

ULF SVANHOLM

Remembering the history of our discipline is refreshing.

BENGT HJALMARSSON

I think you're getting old.

ULF SVANHOLM

What's age got to do with it?

BENGT HJALMARSSON

In science, only the old are into the past.

ULF SVANHOLM

And you?

BENGT HJALMARSSON

I'm interested in my future . . . another reason why I'm now heading for the lab. See you at the next meeting. *(Exit)*

(Women enter from right upstage)

ULLA ZORN

I can't just sit there . . . you've got to tell them.

ASTRID ROSENQVIST

The next time. Satisfied?

ULLA ZORN

Yes. *(Pause)*. May I ask you a question?

ASTRID ROSENQVIST

Certainly.

ULLA ZORN

What do <u>you</u> get out of this?

ASTRID ROSENQVIST

You mean chairing the committee?

ULLA ZORN

Just being on it.

ASTRID ROSENQVIST

Wouldn't you like being both judge and jury? Expecting honors for being first is the occupational disease of scientists. God knows we're not in it for the money. And when we're writing papers, we're supposed to behave like "gentlemen" . . .

(They laugh)
out to pursue knowledge. But Nobel committees are special: we hand out the biggest pat on the back in science –

ULLA ZORN
Without craving it yourself?

ASTRID ROSENQVIST
I didn't say that.

ULLA ZORN
I hope you don't mind my asking: what about you and the Nobel Prize?

ASTRID ROSENQVIST
No Swedish woman has ever gotten it in any science. One will eventually.

ULLA ZORN
Aren't you the first woman who has ever chaired a Nobel Committee?

ASTRID ROSENQVIST
Yes.

ULLA ZORN
How important is that to you ... being first?

ASTRID ROSENQVIST
You are beginning to sound like a district attorney... or a shrink.

ULLA ZORN
Sorry about that. I just wanted to know what price you're willing to pay to be successful as a scientist... and as a woman.

ASTRID ROSENQVIST
I have no children. Many would consider that a heavy price.

ULLA ZORN
Like Mme. Lavoisier? *(Pause)*. Is the committee your child?

ASTRID ROSENQVIST

It certainly doesn't feel or behave like a baby ... but it's a challenge. A contentious committee with a tough job: proposing the first retro-Nobel winner in Chemistry. If we come up with a persuasive choice, the Academy is likely –

ULLA ZORN

... to rubberstamp it?

ASTRID ROSENQVIST

(Laughs)

Those are your words ... not mine. To be persuasive, we also have to be unanimous ... or at least close to it. I have to engineer that. Not an easy job ... you may have sensed some of the crosscurrents.

ULLA ZORN

I did *(pause)* ...

ASTRID ROSENQVIST

Ulf and Sune aren't very subtle.

ULLA ZORN

I meant you and Bengt.

BLACKOUT

SCENE 5

Stockholm, 1777.

SCHEELE *and* MME. LAVOISIER *meet.*

MME. LAVOISIER
Ah . . . Monsieur Scheele! Have you seen my husband? Tonight's masque still requires some preparation.

SCHEELE
I have not. But Madame . . .

MME. LAVOISIER
Yes?

SCHEELE
I understand you keep your husband's correspondence.

MME. LAVOISIER
How did you come to know that?

SCHEELE
Fru Pohl told me.

MME. LAVOISIER
She tells you everything?

SCHEELE
She is an honest woman. She shares the good . . . and the bad with me.

MME. LAVOISIER
Like a wife.

SCHEELE
Or a friend. But, Madame, since you read everything sent to your husband . . .

MME. LAVOISIER
I try.

SCHEELE
One question then.

MME. LAVOISIER
Yes?

SCENE 5

SCHEELE

Sara Margaretha mentioned the letter I dispatched three years ago –

MME. LAVOISIER

(Quickly cries out, while pointing offstage)

Oh . . . there goes Antoine. I must catch him.

LIGHTS DOWN *and then* UP *on* FRU POHL *and* LAVOISIER

FRU POHL

Monsieur Lavoisier! What luck to encounter you . . .

LAVOISIER

Madame will excuse me, but I must prepare for tonight's masque.

FRU POHL

Surely you have time for one simple question?

LAVOISIER

A lady's questions are rarely simple.

FRU POHL

A short one then?

LAVOISIER

Even worse: short questions are never simple.

FRU POHL

Monsieur . . . I'm not clever with words.

LAVOISIER

But you are disarmingly persistent. Your question then? Your one question?

FRU POHL

Yesterday . . . in the sauna–

LAVOISIER

(Quickly)

A curious Nordic custom . . . but one my wife found bracing.

FRU POHL

It was my idea to invite the ladies.

LAVOISIER

Nudity can be disarming.

FRU POHL

Madame Lavoisier was not disarmed.

LAVOISIER

Of course to disarm . . . one first needs to be armed.

FRU POHL

Your wife was.

LAVOISIER

Madame Pohl, you are observant.

FRU POHL

Women from the countryside have to be.

LAVOISIER

Touché, Madame. But your question . . . your simple, short question?

FRU POHL

Why?

LAVOISIER

(Taken aback)

Is that your question?

FRU POHL

Yes.

LAVOISIER

It is indeed short . . . but is it simple? Why what?

FRU POHL

Why did you accept our King's invitation?

LAVOISIER

(Looks at her for a long time)

You are a clever woman, Madame Pohl.

(Prepares to exit)

LIGHTS DOWN *and then* UP
on PRIESTLEY *and* MME. LAVOISIER

PRIESTLEY

Much has happened since we last met in Paris . . .

MME. LAVOISIER

Three years is a long time ...

PRIESTLEY

Only the young would think so ...

MME. LAVOISIER

Are the young not entitled to an opinion?

PRIESTLEY

Opinion? Of course. *(Pause)*. I was referring to Madame's judgment.

MME. LAVOISIER

Perhaps women mature faster in France ...

PRIESTLEY

A conclusion you shared already with my wife ...

MME. LAVOISIER

So she told you about our meeting?

PRIESTLEY

My wife hides nothing from me.

MME. LAVOISIER

(Sotto voce)

That I would call poor judgment.

PRIESTLEY

Why?

MME. LAVOISIER

Some things ought to be hidden ... even in a sauna.

PRIESTLEY

Another opinion ... or another judgment?

MME. LAVOISIER

Solely a comment. But no matter. *(Pause)*. You seem vexed, Monsieur ... I trust I'm not the cause.

PRIESTLEY

Three years ago ...

MME. LAVOISIER

You dined at our table ... content and eager.

PRIESTLEY

You translated ...

MME. LAVOISIER

I tried my best . . . and you seemed grateful.

PRIESTLEY

I was then.

MME. LAVOISIER

But not now?

PRIESTLEY

I'm not sure you transmitted everything . . .

MME. LAVOISIER

Perhaps my knowledge of English is wanting . . .

PRIESTLEY

Madame's English is excellent.

MME. LAVOISIER

I appreciate the compliment. *(Pause)*. Of course, a translator is also a filter, a sieve . . .

PRIESTLEY

Whose efficacy depends on the mesh.

MME. LAVOISIER

Indeed it does . . . and mine is fine.

PRIESTLEY

I am speaking of filtering information . . . not impurities.

MME. LAVOISIER

But so am I, Monsieur.

END OF SCENE

SCENE 6

(A bare room, except for a theatrical curtain. Sitting with their back to audience are Dr. and Mrs. Priestley, Scheele and Fru Pohl. There is a suggestion of a Royal Box, occupied. M. and Mme. Lavoisier enter.)

LAVOISIER and MME. LAVOISIER

(Deep bow, curtsy)

Your Majesties!

LAVOISIER

Dr. and Mrs. Priestley!

MME. LAVOISIER

Apothecary Scheele, Fru Pohl . . .

LAVOISIER

Welcome!

MME. LAVOISIER

Knowing of your love for the stage and opera, Your Majesty . . .

LAVOISIER

In this, your magnificent court theatre at Drottningholm . . .

MME. LAVOISIER

In the tradition of the court of our King Louis the Sixteenth . . .

LAVOISIER

We bring you a small entertainment, a masque, of . . .

MME. LAVOISIER

The Victory of Vital Air . . .

LAVOISIER

Over Phlogiston!

(Music, by Lully, Rameau, Mozart, or possibly the Swedish composer Johan Helmich Roman, rings out majestically. Scheele and Priestley move to show their apprehension. The Lavoisiers, each of them after putting on a mask, part the curtains, whereupon the music should drop off to a low level as the actual masque begins, or change to a recitative harpsichord accompaniment)

LAVOISIER *(playing* PHLOGISTON)

(Broad, even lewd comedy is in order, plus pretentious music. He declaims, preferably in recitative)

I am the vital fire of chemistry,
The element that sets the others free.
The Greek philosophers were unaware
Of how I act on water, earth and air.
Without me, Phlogiston, the world would be
Quite unillumin'd, rudimentary.
'Tis in my gift, the elements to bind,
Transforming them to everything we find:
The precious salts and metals, other earths,
Endowed by me, can offer up their worth.

*(*PRIESTLEY *and* SCHEELE *couples nod approvingly, mime applause)*

MME. LAVOISIER *(playing* OXYGEN)

(Wearing the mask of oxygen)

Monsieur, you are most glittering and sure
Of what the world is made of. Tell me more!
You say there's a terra this, a terra that,
Pray show me how these elements react.

LAVOISIER

A just enquiry, Madam, first, take fire.
All things that burn release me to the air.
Take charcoal, fat, they're full of Phlogiston
And when their blazing ceases, I am gone

MME. LAVOISIER

You have an end?

LAVOISIER

No! Listen carefully!
Air can only hold just so much of me.
There can be other ways I also may appear:
When pure metal rusts, off I fly, my dear!

MME. LAVOISIER

Your miracles are endless! Tell me more!

SCENE 6

LAVOISIER

'Tis I that wins the metal from the ore.
My role in such extraction will astound,
Remember that in charcoal I am found,
And from the coal the ores just suck me out!

MME. LAVOISIER

A marvel, Sir, yet you leave much to doubt –
Your theory is hopelessly behind!
We now know air to be of different kinds:
Inflammable, nitrous, vital and fixed.
And water's not an element, but mixed.
As mon mari *will most ably demonstrate!*

(Priestley becomes very agitated at this point)

LAVOISIER

A revelation, I'll accommodate!

MME. LAVOISIER

My husband soon will show that much depends
On vital air – that he calls Oxygen.
You boast that Phlogiston's the key to fire
And rust, but why not credit vital air?
Could it not feed the flames or lead to rust?
Combined with carbon, say, or iron, it must!
You claim that metal needs you but what for,
When charcoal coaxes Oxygen from ore?
Another point I feign cannot be true
Is your idea of rust. Surely, you knew
That metal grows in weight when thus decayed.
Yet you insist that nothing new is made!

LAVOISIER

(Embarrassed)
My dear, (Pause) . . . Phlogiston might just be so light
That it is weightless. Could he still be right?
(Tentatively, dances with large balloon to gain elevation)

MME. LAVOISIER

Mon cher monsieur, you're speaking like an ass!
You know there's no such thing – negative mass!
A revolution is about to dawn
In chemistry, as Oxygen is born.
Phlogiston is a notion of the past,
Disproved and set aside, indeed, surpassed.

(Priestleys, Scheele, and Fru Pohl grow more agitated from here on to end of scene)

Elements may be many. Some know we,
The rest await skilled hands to set them free.
In chemical reaction, this remains:
Matter is neither lost nor gained.
In this new chemistry, let us rejoice
And thank our sovereign patrons in one voice:
Our Louis, George, and gallant King Gustaf,
In whose light we gather to finish off
Vain Phlogiston. Now join to celebrate
How vital air triumphed in this debate!

(Phlogiston and Vital Air struggle to the concluding music. Mme. Lavoisier pricks balloon with hatpin, whereupon it explodes. Phlogiston sinks to the ground. The Priestleys, Scheele and Fru Pohl overturn their chairs and rush off stage)

*(*LAVOISIER *and* MME. LAVOISIER *drop their masks on the floor)*

LAVOISIER

They weren't amused! Perhaps we went too far.

MME. LAVOISIER

We planted a seed ... their doubt will grow.

LAVOISIER

I worry.

END OF SCENE 6

END OF ACT

SCENE 7

(Stockholm, 2001; Royal Swedish Academy of Sciences, two weeks later than Scene 4).

ASTRID ROSENQVIST

So Scheele on his deathbed marries the widow of the pharmacist who preceded him. Touching, Sune . . . but how relevant is it?

ULF SVANHOLM

(Irritably)

The retro-Nobel will be given for their work . . . not their private lives!

SUNE KALLSTENIUS

What if you can't separate the two?

BENGT HJALMARSSON

Lavoisier certainly had a private life! He even got his head chopped off . . . and that had nothing to do with his chemistry. He was a tax collector . . . hardly a popular occupation during the French Revolution. *(Pause)*. But did your man Scheele and Fru Pohl live together?

SUNE KALLSTENIUS

It depends on one's definition of "living together." For most of the time, they occupied the same house, which she kept for Scheele. *(Pause)*. But did they cohabitate? It's been said of Scheele "that he never touched a body without making a discovery." But these bodies were chemicals, not women. In my opinion, Scheele was celibate all his life . . . a chemical monk.

ULLA ZORN

Very clever!

SUNE KALLSTENIUS

Ms. Zorn . . . you sound as if you have an opinion on this topic. After all, your're the one who mentioned wives.

ULLA ZORN

(Quick, but low voice)

Yes.

SUNE KALLSTENIUS

Yes . . . you have special knowledge . . . or yes, they did?

ULLA ZORN

"Yes" to the former . . . and "perhaps" to the latter.

LIGHTS FADE on committee.

*(*ROSENQVIST, HJALMARSSON *and* SVANHOLM *freeze, while* ULLA ZORN *and* SUNE KALLSTENIUS *change costumes on stage and move downstage and across).*

LIGHTS UP on SCHEELE and FRU POHL

*(*FRU POHL *crosses over to a sideboard and mimes the grinding of coffee)*

FRU POHL

Carl Wilhelm . . . it's time you came in. It's so cold in the shed. If only you could afford a proper laboratory.

SCHEELE

(Stamping his feet)

I know you care for me, Sara. But it took time to dissolve in acid this ore Bergman sent to me. There may be a new metal in it.

FRU POHL

My son and I have eaten. But there is food on the table for you. *(She hesitates)*. And a letter from the printer, Swederus, in Uppsala.

SCHEELE

No book?

FRU POHL

He promises it.

SCHEELE

(Exasperated)

But when? I finished it last year. He sat on it for

months. I complained. Now three more months have passed, and my experiments on fire air gather dust in that damned printer's shop.

FRU POHL

Be patient, others know of your work.

SCHEELE

A few friends ... in Sweden. The book will go far beyond our borders.

FRU POHL

I would help you, Carl Wilhelm. If only I were not so ignorant ...

SCHEELE

Your concern is more important to me. But now I must finish that letter.

FRU POHL

To whom?

SCHEELE

Monsieur Lavoisier, the French chemist. He has burning lenses, Sara, that are as big as our house.

FRU POHL

Monsieur Lavoisier must really want to set things on fire.

SCHEELE

His is hotter. He can make chemical reactions go as no one else can.

FRU POHL

But can he turn a profit from a pharmacy?

SCHEELE

He might, for he also knows the way of money. *(Pause)*. In my letter I ask him to repeat my experiments making fire air.

FRU POHL

Why ask him?

SCHEELE

Because my air is new. Because there is no better way of having the world know than asking one of the best scientists to repeat my experiment.

FRU POHL

(Hesitatingly)

Forgive my forwardness, Carl Wilhelm . . . but . . . is that what you desire most? That the world knows of you?

SCHEELE

(Taken aback)

No one has asked me that before. *(Reflects)*. Respect is important –

FRU POHL

You have it from the citizens of Köping.

SCHEELE

I want to be my own master, a simple thing. And I want to earn enough money . . . to support you and your son –

FRU POHL

We manage.

SCHEELE

Because of your frugality.

FRU POHL

I've never complained.

SCHEELE

I know . . . I want to earn enough to buy better supplies, a more powerful burning lens–

FRU POHL

And to heat your laboratory! Carl Wilhelm . . . I worry about your health.

SCHEELE

(Moved, takes her hand, pauses to inspect his hand and then hers)

Look! The coffee sticking to your hand! Is it some form of magnetism?

LIGHTS FADE

(SCHEELE and FRU POHL change costumes on stage and then rejoin other committee members)

ULLA ZORN

You see? He did touch part of her body and made a discovery. *(Pause)*. And that could have been personal magnetism.

BENGT HJALMARSSON

(Astonished)

Where did you dig that up?

ULLA ZORN

Scheele mentioned that incident in a letter to Johan Carl Wilcke, the secretary of the Royal Swedish Academy of Sciences.

SUNE KALLSTENIUS

But how did you come across that letter?

ASTRID ROSENQVIST

(Interrupts)

Later.

BENGT HJALMARSSON

No, Astrid! Not later! Now!

ASTRID ROSENQVIST

What's the urgency?

BENGT HJALMARSSON

I have a feeling your job description of "amanuensis" bears no relation to the dictionary definition. *(Turns to Ulla Zorn)*. Where did you discover those nuggets of information?

ASTRID ROSENQVIST

All right, Ulla ... tell him.

ULLA ZORN

I'm finishing my Ph. D. at Lund University ...

SUNE KALLSTENIUS

These days, most chemistry students barely know who Scheele was.

ULLA ZORN

Perhaps a reflection on the professors rather than the students.

ULF SVANHOLM

Touché.

BENGT HJALMARSSON

(To Ulf Svanholm, sarcastically)

I see you haven't forgotten your high school French.

(To Ulla Zorn)

But that letter to Wilcke where Scheele talks about grinding coffee with his girlfriend or whatever she was ... where did you run into that?

ULLA ZORN

Her name was Sara Margaretha Pohl. And I found it the same way you would have: research!

BENGT HJALMARSSON

(Ironic)

I see.

(Shifts to ordinary tone).

In that case, let me tell you about my research ... But before telling you about Lavoisier, the chemist, banker and economist ... who did everything from debunking mesmerism to shipping gunpowder to the Americans ... listen to some goodies about Madame Lavoisier

ULLA ZORN

My, my! I never thought my comments about wives would have such an effect on this committee.

BENGT HJALMARSSON

Don't flatter yourself, Ms. Zorn. I always throw a wide net in my research.

ASTRID ROSENQVIST

Especially when it comes to women! *(Laughs)*. Sorry about that ... it just slipped out. Go ahead, Bengt ... tell us what you caught in your net.

BENGT HJALMARSSON

First of all, Madame Lavoisier wasn't just his wife ...

(Mockingly to Ulla Zorn)

she was his amanuensis ... Of course not full time.

ULLA ZORN

(Coldly)

It's not a very attractive full time position ... for an ambitious woman.

ASTRID ROSENQVIST

Everything is possible for an ambitious woman ...

BENGT HJALMARSSON

She even helped in the lab ... Yet she was barely in her teens when she married Lavoisier... her first husband.

ULF SVANHOLM

First husband? How many men were there?

BENGT HJALMARSSON

The second husband, Count Rumford, I think she would have liked to forget ... even though he was almost as famous as Lavoisier. But men? Probably a fair number... even by present day standards. Benjamin Franklin was quite smitten by her. But Pierre Samuel Du Pont ...

SUNE KALLSTENIUS

The American Du Pont? The millionaire chemist?

BENGT HJALMARSSON

His French father. Now that was a different story, a love story.

(Picks up a paper)

Du Pont wrote her a letter four years after Lavoisier's death ... after... I quote from his letter... "twenty-two years of acquaintance and seventeen of intimacy."

(Pause).

In other words, they had been "intimate"... for at least thirteen years while the Lavoisiers were still married.

ASTRID ROSENQVIST

A modern couple ...

BENGT HJALMARSSON

(Continues reading, but facing Rosenqvist as if words were meant for her).

"If you could have continued to love me, I would have patiently accepted that destiny . . ."

(Looks up from letter, addresses Rosenqvist)

That's Du Pont speaking . . . not me . . .

(Again picks up letter and continues to read)

"because a single evening with you around the fire . . . would have been compensation for both my eyes and heart . . . I belonged to you, my dear young lady . . ."

The young lady was then forty-one!

(The ringing of a cellular phone is heard. The committee members, startled, look around, perhaps also toward the theatre audience as if the phone might come from there).

ULLA ZORN

(Flustered, rummages through handbag, while the phone keeps ringing, perhaps with an annoying musical tone. She finally retrieves phone and whispers audibly)

Yes? *(Brief pause)*. To Ithaca. *(Brief pause)*. New York . . . *(Brief pause)*. Economy! *(Brief pause)*. Three days only . . . maximum four. *(Brief pause)*. Call later . . . I can't talk now.

(Puts down phone. Looks unapologetic)

Sorry . . . I didn't know it was on.

LIGHTS DIM.

(BENGT *and* ULF *move to one side of stage,*
ASTRID *and* ULLA *to other. Sune remains frozen in his seat.*
Spotlight on faces of BENGT *and* ULF).

BENGT HJALMARSSON

That telephone call.

ULF SVANHOLM

I won't touch a cellular phone.

BENGT HJALMARSSON

Another sign you're getting old. *(Laughs)*. . . Why is she flying to Ithaca?

ULF SVANHOLM

Probably a boyfriend . . . at Cornell University.

BENGT HJALMARSSON

I doubt it.

CROSSFADE to spotlight on faces of women.

ULLA ZORN

You aren't angry, are you?

ASTRID ROSENQVIST

Just amused.

ULLA ZORN

That's a relief.

ASTRID ROSENQVIST

But you're showing off too much.

ULLA ZORN

Bengt Hjalmarsson irritates me.

ASTRID ROSENQVIST

Bengt is a complicated man.

ULLA ZORN

I suppose that's a compliment.

ASTRID ROSENQVIST

Not necessarily . . . it's just an experimental observation

LIGHTS FADE,
rise on committee back in place

ASTRID ROSENQVIST

Ulf . . . since we're hearing about everybody's historical research, what did you dig up on Priestley? Or did you spend your time on Mrs. Priestly?

ULF SVANHOLM

I did not! Priestley lived at the right time in the right country: England . . . the 18th century hot house of pneumatic chemistry. In the case of Priestley, the self-taught chemist just happened to be a minister. He published 50 works on theology, 13 on education, 18 on political, social, and metaphysical subjects . . .

BENGT HJALMARSSON

A preacher dabbling in chemistry . . .

ULF SVANHOLM

(Raises hand)

. . . and fifty papers and no less than twelve books on science! You wouldn't call that dabbling, would you?

SUNE KALLSTENIUS

But what's in those books and papers? We must deal with content . . . with quality . . . not authorial diarrhea!

ULF SVANHOLM

Now, now! Just because Scheele completed only one book . . . just because your man was constipated . . .

ASTRID ROSENQVIST

(Admonishingly)

Enough! What about the chemistry?

BENGT HJALMARSSON

Did Priestley realize what he was doing?

ULF SVANHOLM

He subjected air to all kinds of chemical processes . . .

BENGT HJALMARSSON

In an utterly haphazard manner.

ULF SVANHOLM

(Beginning to show irritation)

He was learning as he went along. When Lavoisier made his "vital air" he used Priestley's method, didn't he? It's the results that count. And–in contrast to Scheele – Priestley was ambitious enough to let people know what he found.

SUNE KALLSTENIUS

Maybe that ambition clouded his judgment.

ULF SVANHOLM

What's wrong with ambition? Look at it as the blemish in a Persian carpet that makes it valuable.

SUNE KALLSTENIUS

Does that mean that a carpet without blemishes cannot be as valuable . . . or even more so?

ULF SVANHOLM

I'm beginning to regret having mentioned ambition . . . or carpets. Let's forget about both! Anyway . . . Priestley loved to talk of his work . . . probably even to his wife.

(Ironic tone)

Or does that surprise you, Ms. Zorn?

ULLA ZORN

Why should it? Mrs. Priestley was educated . . . she wrote beautiful letters . . . and she was a true helpmate.

LIGHTS FADE on committee.

*(*HJALMARSSON, ZORN *and* KALLSTENIUS *freeze,*
while ASTRID ROSENQVIST *and* ULF SVANHOLM *change costumes on stage and move downstage and across)*

LIGHTS UP on PRIESTLEY and MRS. PRIESTLEY.

MRS. PRIESTLEY

And what did you do in Paris?

PRIESTLEY

I visited Versailles with Lord Shelburne.

MRS. PRIESTLEY

And dined well, no doubt.

PRIESTLEY

In fact, one night at the table of Monsieur and Madame Lavoisier. Most of the natural philosophers of the city were there. I told them of my new air, in which a candle burned much better than in common air.

MRS. PRIESTLEY

I wish you had taken me, Joseph.

PRIESTLEY

I wish you had been there. Oh, it was difficult, Mary.

MRS. PRIESTLEY

They did not believe you?

PRIESTLEY

Who knows? I spoke French so imperfectly – I had common words, but not the scientific terms.

MRS. PRIESTLEY

I would have translated for you.

PRIESTLEY

You're a clever woman, Mary... but what about the children? No matter... Mme. Lavoisier asked how I made the air.

MRS. PRIESTLEY

(Concerned)

And you told her?

PRIESTLEY

Of course. Mme. Lavoisier understood. She explained it to her husband.

MRS. PRIESTLEY

She also assists him in the laboratory?

PRIESTLEY

Indeed. After dinner, she displayed drawings of their equipment... much better than mine... which, I hope, will persuade Lord Shelburne to loosen his purse some more. But her drawings were skillful...

MRS. PRIESTLEY

I envy her. I once learned how to draw...

PRIESTLEY

You help in other ways... you take splendid care of house and family...

MRS. PRIESTLEY

And money. But I worry about its source.

PRIESTLEY

I am dependent on his Lordship's favors...

MRS. PRIESTLEY

Which could be withdrawn without notice.

(She pauses, pointing to a newspaper)

Joseph... have you heard what Samuel Johnson says of you?

PRIESTLEY

That scribbler!

MRS. PRIESTLEY

He says you are "an evil man. His work unsettles everything."

PRIESTLEY

(Laughs)

Not as good as Edmund Burke, Mary. Not nearly so. Burke called me "the wild gas, the fixed air is plainly broke loose."

(Laughs again)

At least he's got one of my airs right!

MRS. PRIESTLEY

I wish you'd be more cautious.

PRIESTLEY

Change will come . . . liberating all the powers of man from the variety of fetters by which they have hitherto been held. Why be afraid? And of whom? Of these arse-lickers of kings?

MRS. PRIESTLEY

But what about your laboratory . . . your work . . . our children? People stir against us in town.

PRIESTLEY

Let them talk. Now let me show you this letter from Dr. Benjamin Franklin . . . His friendship makes up for all the fulminations of the others.

LIGHTS FADE

(PRIESTLEY and Mrs. PRIESTLEY change costumes on stage and then rejoin other committee members)

LIGHTS UP

ULF SVANHOLM

Isn't it ironic? Priestley–a chemical conservative . . . just think of his undying defense of phlogiston–was such a political and religious revolutionary that a mob burned his house in Birmingham. *(Pause)*. Three years later, he fled to America . . . with Benjamin Franklin's help.

ASTRID ROSENQVIST

Can we finally we turn to Scheele's letter? Did Lavoisier get it? Did he read it?

BENGT HJALMARSSON

There is no hint on the part of Lavoisier–no letters, hearsay, other contemporary documents . . . at least in France . . . absolutely nothing to indicate that he had ever received a written communication from Scheele. Yet the answer is . . . *(pause)*. . . yes, he did.

ASTRID ROSENQVIST

Yes to both questions?

BENGT HJALMARSSON

To both.

ULLA ZORN

And the evidence?

BENGT HJALMARSSON

Grimaux's finding.

ULLA ZORN

(Dismissive, but sotto voce)

Oh that!

ASTRID ROSENQVIST

Who is Grimaux?

BENGT HJALMARSSON

A French chemist turned historian who found Scheele's letter in 1890. There it was . . . hidden among Lavoisier's papers for over one hundred years.

ASTRID ROSENQVIST

And you saw it?

BENGT HJALMARSSON

(Starts rummaging in his briefcase)

Yes. It's now in the archives at the French Academy of Sciences. *(Triumphant)*. And I brought some slides to prove it. Here is the first page:

(Projects Fig. 1a and then walks up to the screen to point to relevant sentence. Reads it quickly in French).

"Je ne desire rien avec tant d'ardeur que de vous pouvoir faire montrer ma reconnaissance." Not bad, eh?

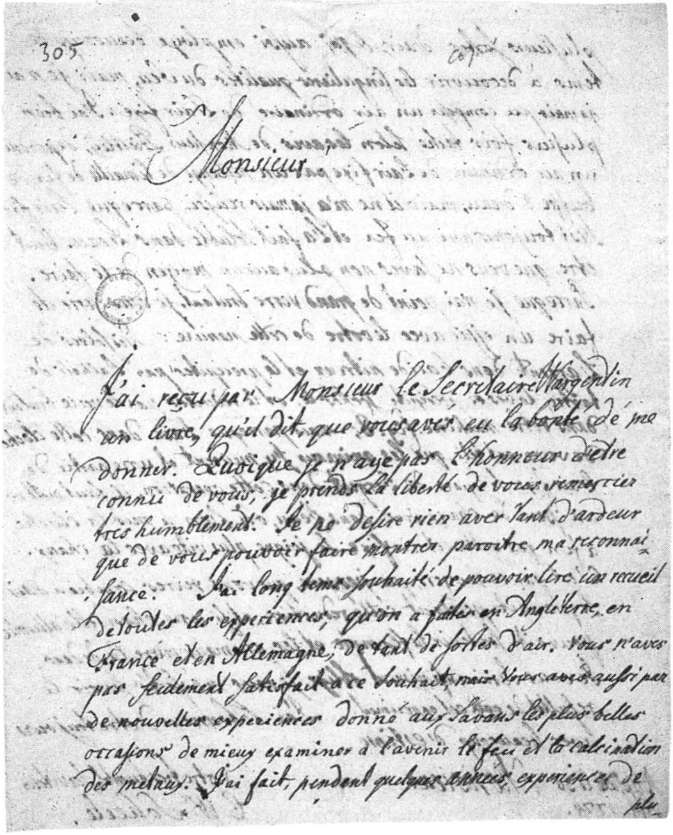

305

Monsieur

J'ai reçu par Monsieur le Secretaire Wargentin son livre, qu'il dit, que vous avez eu la bonté de me donner. Quoique je n'aye pas l'honneur d'etre connu de vous, je prends la liberté de vous remercier tres humblement. Je ne desire rien avec tant d'ardeur que de vous pouvoir faire montrer paroitre ma reconnaissance. J'ai long tems souhaité de pouvoir lire un recueil de toutes les experiences, qu'on a faites en Angleterre, en France et en Allemagne, de tant de sortes d'air. Vous n'avez pas seulement satisfait a ce souhait, mais vous avez aussi par de nouvelles experiences donné aux savans les plus belles occasions de mieux examiner à l'avenir le feu et la calcination des metaux. J'ai fait, pendant quelques années, experiences de plu-

Fig. 1a

SUNE KALLSTENIUS

Stop showing off and translate.

BENGT HJALMARSSON

"There is nothing I desire more eagerly than to be able to show you my discovery."

SUNE KALLSTENIUS

(Grins)

Well, well! So you're moving to the Scheele camp!

ULLA ZORN

Professor Hjalmarsson . . . I hope you won't mind a minor correction.

BENGT HJALMARSSON

What correction?

ULLA ZORN

"*Reconnaissance*" means "gratitude," not "discovery." Scheele is just thanking Lavoisier for a book he had sent him earlier.

BENGT HJALMARSSON

(Slightly irritated, but recovers quickly)

Of course! How stupid of me. But where did you pick up your French?

ULLA ZORN

A former boyfriend . . . He was French.

BENGT HJALMARSSON

Ah, yes . . . by far the most efficient way of learning French . . . But back to the letter. Here is the second slide:

(Projects Fig. 1b)

SUNE KALLSTENIUS

(Jumps up, goes to screen and points to bottom lines)

Note the date, September 30, 1774. And Scheele's signature.

ULLA ZORN

But that doesn't prove that Lavoisier had actually read the letter.

(All look at her, startled)

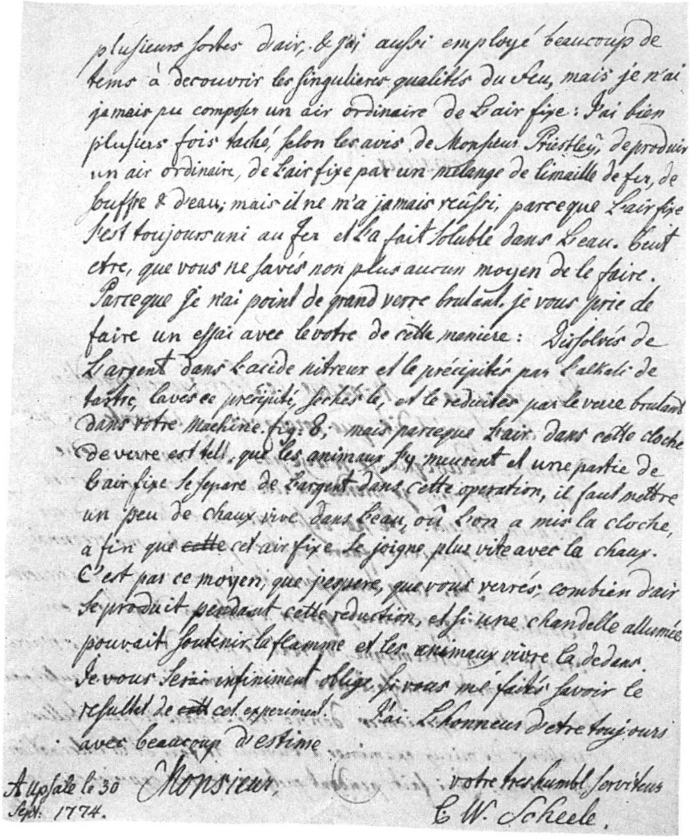

plusieurs sortes d'air, & j'ai aussi employé beaucoup de
tems à decouvrir les singulieres qualités du Feu, mais je n'ai
jamais pu composer un air ordinaire de L'air fixe: J'ai bien
plusieurs fois taché, selon les avis de Monsieur Priestley, de produir
un air ordinaire, de L'air fixe par un melange de limaille de fer, de
souffre & d'eau; mais il ne m'a jamais reussi, parceque L'air fixe
s'est toujours uni au fer et L'a fait soluble dans L'eau. Peut
etre, que vous ne savés non plus aucun moyen de le faire.
Parceque je n'ai point de grand verre brulant, je vous prie de
faire un essai avec le votre de cette maniere: Dissolvés de
L'argent dans L'acide nitreux et le précipités par L'alkali de
tartre, lavés ce precipité, seches le, et le reduisés par le verre brulant
dans votre Machine. fig: 8; mais parceque L'air dans cette cloche
de verre est tell, que les animaux s'y meurent et une partie de
L'air fixe se separe de L'argent dans cette operation, il faut mettre
un peu de chaux vive dans L'eau, où L'on a mis la cloche,
a fin que ~~cette~~ cet air fixe se joigne plus vite avec la chaux.
C'est par ce moyen, que j'espere, que vous verrés, combien d'air
se produit pendant cette reduction, et si une chandelle allumée
pouvait soutenir la flamme et les animaux vivre la dedans.
Je vous serai infiniment obligé, si vous me faités savoir le
resultat de ~~cett~~ cet experiment. J'ai L'honneur d'etre toujours
avec beaucoup d'estime

A Upsale le 30 Sept: 1774.

Monsieur,

votre tres humbl serviteur
C. W. Scheele.

Fig. 1b

BENGT HJALMARSSON

And what makes you say that?

ULLA ZORN

A historian's skepticism.

BENGT HJALMARSSON

(Taken aback, walks over to Zorn to confront her)

What did you say your Ph. D. thesis was on?

ULLA ZORN

I didn't.

BENGT HJALMARSSON

Some national secret that you cannot share with us?

ULLA ZORN

Not at all. You just never asked me.

BENGT HJALMARSSON

I am asking now.

ULLA ZORN

"Women in the lives of some 18th century chemists."

BENGT HJALMARSSON

And why didn't you tell us that before?

ULLA ZORN

You didn't seem to have a very high opinion of historians . . . Perhaps you still don't.

END OF SCENE 7

SCENE 8

(Stockholm 2001, immediately after Scene 7. Committee members other than Hjalmarsson have left, while Zorn is collecting papers and closing computer).

BENGT HJALMARSSON

Now that everyone has left, I hope you won't mind a comment.

ULLA ZORN

I had no choice before . . . so one more won't hurt me.

BENGT HJALMARSSON

So Astrid felt she had to smuggle you in? Don't you feel used?

ULLA ZORN

With you, I might have. But not with Professor Rosenqvist.

BENGT HJALMARSSON

"Professor Rosenqvist!" Why don't you call her "Astrid?"

ULLA ZORN

I do . . . usually.

BENGT HJALMARSSON

Why not now?

ULLA ZORN

Out of respect . . . for her. I didn't like the way you questioned her . . . about my presence.

(Hjalmarsson looks at her, then sits on edge of table facing her)

BENGT HJALMARSSON

You're right . . . I was irritated. I don't like to be surprised. Of course, you didn't behave like a mousy amanuensis. *(Laughs)*. What a precious word . . . "amanuensis."

ULLA ZORN

For once we agree. I hate it. Nor do I enjoy being reduced to a keeper of committee minutes . . . even of a Nobel Committee.

BENGT HJALMARSSON

You certainly showed it.

ULLA ZORN

(Sarcastic)

I'll try to improve . . . now that you know I am also a historian. Imagine how I felt at the first meeting when you all disparaged my profession–

BENGT HJALMARSSON

How could we've known that a historian was in the room?

ULLA ZORN

You wouldn't have acted differently.

BENGT HJALMARSSON

(Laughs)

Probably not.

ULLA ZORN

(Musingly).

Actually, I have a feeling something has changed.

BENGT HJALMARSSON

Yes?

ULLA ZORN

Now you all think you're experts in <u>my</u> field.

BENGT HJALMARSSON

Is that why you brought in those historical tidbits? To show that we still had to learn?

ULLA ZORN

(Changes the topic)

When I see all of you . . . sniping at each other . . . worrying about who published . . . who didn't . . .

BENGT HJALMARSSON

You're puzzled.

ULLA ZORN

This wasn't my idea of science and scientists.

BENGT HJALMARSSON

You think we arrange beetles in a museum case?

SCENE 8

ULLA ZORN

I thought at the heart of science was sheer curiosity. I see that in Scheele . . . maybe also in Priestley. I start having troubles with Lavoisier.

LIGHTS FADE

(Trumpets, as at the Nobel Award ceremonies, signaling a change of place and time.)

LIGHTS UP

(A suggestion of a palace setting, a royal theatre. At center or right upstage is a bare demonstration table. Actual or simulated experiments will be done at this table; there may also be projections shown on rear screen. Left downstage is the sauna where the women will appear.)

COURT HERALD'S VOICE

Your Majesties, esteemed guests! Throughout Europe, pneumatic chemistry is in the air. A dispute has arisen: Who, among these great savants, discovered the vital air supporting life? *(Pause)*. A golden medal . . . with a likeness of our King Gustavus III . . . will be struck in honor of the true discoverer. And our King is famed for his munificence in other ways . . .

PRIESTLEY

(Aside)

As he squanders the people's money . . .

(Trumpets)

COURT HERALD'S VOICE

Let the Judgement of Stockholm begin! And let the three savants be their own judges! Vital air! *(Pause)*. Who made it first?

SCHEELE

(Quietly, but quickly)

I did. And called it *eldsluft* . . . a good Swedish word for fire air.

PRIESTLEY

But is that not air deprived of all phlogiston? The air that inflames all things? That is why I named it "dephlogisticated air." *(Pause)*. But dear Scheele ... where should we have learned of your discovery?

SCHEELE

In my book, about to appear ...

PRIESTLEY

I made that air by heating *mercurius calcinatus* in 1774 and ...

(Pause, then raised voice for emphasis)

communicated that discovery in the same year!

(Addresses Scheele)

I know of no paper of yours ...

LAVOISIER

(Smiling)

Mes amis! He who starts the hare, does not always catch it.

SCHEELE

There is no hare to catch if someone does not start the hunt!

LAVOISIER

It is we who must decide who first captured the essence of that vital air ...

PRIESTLEY

(Sarcastic)

And what does that mean?

SCHEELE

It is essential to know who made the air first ...

PRIESTLEY

... for it is the invention that will be remembered by posterity, not its ephemeral interpretation ...

LAVOISIER

(Shifting the subject)

Let us do the experiments we judge vital in this matter. Whose experiment will come first?

SCENE 8

SCHEELE

Monsieur Lavoisier, do me the honor of performing the experiment I brought to your attention some three years ago in my letter–

LAVOISIER

I know of no letter–

SCHEELE

(Takes paper from his coat)

Let me read it for you.

*(*LIGHTS DIM*; spots on two men. This is the first of three experimental scenes. The stage is darkened, except for spots on the bench and on the man who performs the experiment, as well as the one who directs him.)*

SCHEELE

(Reads from letter in his hand).

Dissolve silver in acid of nitre and precipitate it with alkali of tartar. Wash the precipitate, dry it, and reduce it by means of a burning lens ... A mixture of two airs will be emitted. And pure silver left behind.

LAVOISIER

And then?

*(*LIGHT DOWN *on men, who continue their experiment in mime.* LIGHT UP *on women in sauna).*

MRS. PRIESTLEY

How hot the air is in your sauna, Fru Pohl!

FRU POHL

Still ... it's the air we all breathe.

MME. LAVOISIER

Yet only part of it is vital air, the rest ...

FRU POHL

Indeed, this part is the air Apothecary Scheele made. Once he invited me into his shed to show me the experiment making fire air he had done earlier in Uppsala. He was bubbling the newly formed gas through a kind of water.

MME. LAVOISIER

It must have been limewater.

MRS. PRIESTLEY

It turned cloudy, didn't it?

FRU POHL

How do you know?

MRS. PRIESTLEY

I've listened to Joseph's lectures on fixed air.

MME. LAVOISIER

The same air we expire ... the one we remove by passage through limewater.

FRU POHL

In the remaining air, he bid me thrust a splint that had blown out. Just a glow of a coal at its end. It was toward evening.

(The flaring up of the splint in the men's experiment coincides with its mention by Mrs. Priestley.)

MRS. PRIESTLEY

And it flared in brightest flame ... and kept burning!

FRU POHL

How could you know that?

MRS. PRIESTLEY

Because my Joseph did that too.

MME. LAVOISIER

We all did it.

(LIGHTS OUT on women, UP on men)

SCHEELE

I did that experiment in 1771 in a pharmacy in Uppsala ... with equipment much more modest than now put at our disposal by your Majesty.

PRIESTLEY

Yet you did not report it?

SCHEELE

I told Professor Bergman ... I thought he would tell others.

PRIESTLEY

Your experiment was with a silver salt.

SCHEELE

I obtained the air over the next three years in many different ways. Including red *mercurius calcinatus*, as you did.

LAVOISIER

That red mercury compound – it it is also how we ... Dr. Priestley and I ... made that air.

PRIESTLEY

We? *(Pause)*. We were not in the same laboratory, Monsieur Lavoisier! Pray speak clearly of who did what and when. I made that air first ... and did so alone. · And I will now show you how I accomplished that. Mr. Scheele, will you perform the experiment?

SCHEELE

It will be an honor to do so.

(Both men step to demonstration table; LIGHTS DIM*)*

PRIESTLEY

In August of 1774, I exposed *mercurius calcinatus* ... the red crust that forms as mercury is heated in air ... in my laboratory to the light of my burning lens. As the red solid is heated, an air will be emitted, while dark mercury globules will condense on the walls of the vessel. You will collect the air by bubbling it through water.

LAVOISIER

But where is your balance, Dr. Priestley? Shall the gas not be weighed?

PRIESTLEY

A timepiece is sufficient. We have here two chambers ... one with ordinary air ... the other with my new dephlogisticated one. Mr. Scheele, now take a mouse ...

*(*LIGHTS DOWN *on the men, who continue to mime experiment.* LIGHTS UP *on women)*

MRS. PRIESTLEY

I asked him–why mice?

FRU POHL

And?

MRS. PRIESTLEY

The good doctor said: Would you use English children? Mice live as we do.

MME. LAVOISIER

On a part of ordinary air.

MRS. PRIESTLEY

Then he placed one mouse in a jar of plain air.

FRU POHL

Where it died in time.

MRS. PRIESTLEY

How do you know that?

FRU POHL

Apothecary Scheele showed me.

MME. LAVOISIER

It is a well-known fact, described also by other savants.

MRS. PRIESTLEY

And then he placed the other one in–

FRU POHL

Fire air . . .

MRS. PRIESTLEY

My Joseph's dephlogisticated air . . .

MME. LAVOISIER

And it lived much longer, did it not? This is why we called that new air eminently respirable. Or vital.

MRS. PRIESTLEY

I detest mice.

FRU POHL

(Laughs)

With living things, Carl Wilhelm can be clumsy. He often dropped them! But we know mice in the country. If I didn't catch them, the cats did.

(LIGHTS OUT on the women, UP on the men)

LAVOISIER

There is no doubt that Dr. Priestley's method produces vital air. But–

PRIESTLEY

But, Monsieur?

LAVOISIER

Now is my turn. May I proceed?

SCHEELE, PRIESTLEY

Of course.

LAVOISIER

We just observed a mouse living longer in the vital air we have all made. Yet in the end that mouse also dies, as the air is depleted. However, in my own work ... I have moved far, far beyond watching mice die. Your Majesty, gentlemen! This air ... which I propose we henceforth call oxygen–

PRIESTLEY

(Interrupts)

I object, sir! It's easy to call something by a new name ... when you don't know what you have! Be descriptive, sir! Why not dephlogisticated–

LAVIOISIER

(Interrupts)

I know the air as well as you do, Monsieur. "Oxy" is Greek ... for sharp or acid. And since I believe our air to be found in all acids, I am being descriptive ...

PRIESTLEY

Descriptive? Bah! You, sir, are being sharp ... or perhaps acid ... but our dephlogisticated air is neither.

LAVOISIER

Allow me the courtesy to continue. This air is at the heart of all chemistry. I have shown that when we breathe, the wondrous machinery of the body transforms a given weight of oxygen ... into other gases and water.

PRIESTLEY

But that is obvious!

LAVOISIER

Not until you weigh it! For that ... *(confronts Priestley)*... a timepiece is not sufficient ... Since nothing is gained ... nor lost in this world ... be it in the economy of a country or a chemical reaction ... the balance sheet of life's chemistry must be determined.

PRIESTLEY

Ah, it's the banker in you ...

LAVOISIER

(Ignores Priestley's comment)

I have brought from Paris a suit of rubber I have devised. It catches all the effluents of the body... to show us that the equation balances. *(Pause)*. Dr. Priestley, are you prepared to perform the experiment?

(LIGHTS OUT, except for spots on Priestley and Lavoisier)

PRIESTLEY

Indeed I am ... even weighing things on your balances. But ... it appears we require a volunteer for the experiment ... to wear your modern suit of armor. Mr. Scheele?

SCHEELE

With pleasure.

(Scheele marches up, with determination. He picks up "rubber suit", not unlike old-fashioned diving or scuba suit).

LAVOISIER

Not only must you weigh Apothecary Scheele ... you must also weigh his suit. The measurements will take several hours.

LIGHTS DIM *on men.*

SCENE 8

MME. LAVOISIER

Ladies . . . I would show you the sketch of the experiments M. Lavoisier performed.

Projection of one of Mme. Lavoisier's drawings of the experiment appears on screen for remainder of their conversation.

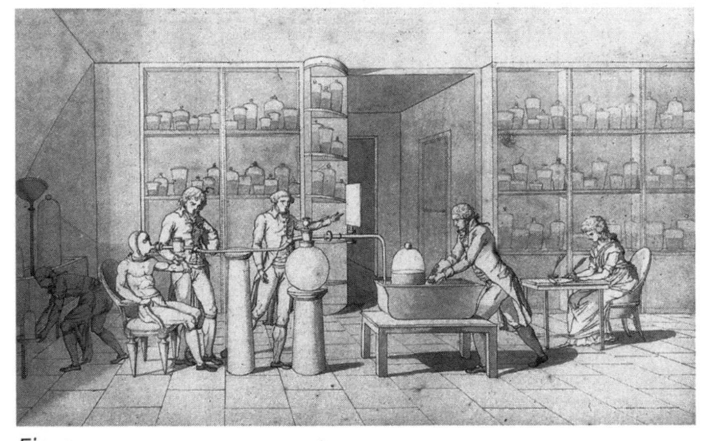

Fig. 2

FRU POHL

A sketch?

MRS. PRIESTLEY

For your own pleasure, Madame?

MME. LAVOISIER

As a record.

FRU POHL

But why should a "record" be needed?

MME. LAVOISIER

To give others evidence of what was done, of course.

MRS. PRIESTLEY

As well as <u>when</u>, I should think.

MME. LAVOISIER

(Startled for a moment)

Our experiments are quite complex. An assistant is encased in a suit of rubber and silk taffeta. And all that goes in and out of him is analyzed. And recorded. Over many hours.

FRU POHL

The poor man!

MME. LAVOISIER

Quantitative analysis is a hard mistress.

(LIGHTS DIM on women, RISE ON men)

LAVOISIER

(Addresses Priestley)

I trust you took care ... for the margin of error must not be more than 18 grains in 125 pounds. What do you find?

PRIESTLEY

Mr. Scheele has lost some weight.

(Scheele seems weak, but smiles)

When we take into account the water breathed out, there's indeed a rough balance.

LAVOISIER

Nothing is created–

PRIESTLEY

Except by God.

LAVOISIER

Nor lost.

SCHEELE

Except by Man. Especially when he is the subject of an experiment.

LAVOISIER

(Driving his point home, and refusing to enter the banter)

Gentlemen! That crucial mass balance *(with emphasis)* ... punctures phlogiston's balloon.

SCHEELE

Surely the facts may be explained otherwise.

PRIESTLEY

Indeed, sir . . .

(He looks at Lavoisier)

. . . the experiment you so laboriously had us do . . . did demonstrate . . . I readily confess . . . one function of your . . .

(Assumes sarcastic tone)

"eminently breathable air." *(Pause)*. But, Monsieur, you did not show us how you made that air.

LAVOISIER

I knew my air was there in ordinary air . . . Did I not see metals combine with it . . . with sulfur . . . or with phosphorus?

PRIESTLEY

That does not tell us how you produced the dephlogisticated air . . .

LAVOISIER

Pray stop calling it "dephlogisticated," Dr. Priestley. The name derives from a theory that is *passé*.

PRIESTLEY

Not for me.

SCHEELE

Nor for me.

LAVOISIER

Why not a new name for the air, to avoid this argument?

PRIESTLEY

Call it oxygen? And yield to the tyranny of a nomenclature invented by you?

LAVOISIER

(Angry)

When a new structure is needed for a science . . . when, indeed, there must be a revolution, new names are also required.

PRIESTLEY

But you did not know what that gas was!

LAVOISIER

I saw the need for one air explaining rusting, burning, and respiration!

PRIESTLEY

(Heatedly)

But until that October dinner in Paris when I informed you of my observations ... you did not know the nature of that air ...

SCHEELE

(Untypically forceful)

And until that October day when you got my letter which told you how to make fire air ...

(They argue simultaneously to the end of the scene)

LAVOISIER

I had begun my experiments with *mercurius calcinatus* ...

PRIESTLEY

Only after you heard of what I discovered ...

SCHEELE

You did not know how to make that air ...

COURT HERALD' VOICE

(Sound of tapping staff)

Gentlemen! Gentlemen! His Majesty is vexed. *(Pause)*. Royal displeasure is the only judgment you will receive today!

(LIGHTS FADE, UP AGAIN IN 2001)

BENGT HJALMARSSON

So Lavoisier is too meticulous for you. Is his interest in precise weights any different from yours in exact dates and documents?

ULLA ZORN

I'm also talking about each of you ...

BENGT HJALMARSSON

You're confusing science with scientists.

ULLA ZORN

Am I?

BENGT HJALMARSSON

Science is a system . . . a search driven by curiosity, all the time touching base with what's real . . . That system works . . .

ULLA ZORN

No matter what motivates the people who do it?

BENGT HJALMARSSON

Scientists might be after priority . . . power . . . money . . . As long as they publish, Ulla, someone will check their work.

ULLA ZORN

And how often does that happen?

BENGT HJALMARSSON

The more interesting the discovery, the more closely it will be checked . . .

ULLA ZORN

To prove the other person wrong? Hardly a noble reason!

BENGT HJALMARSSON

It still keeps us honest . . . most of the time. It doesn't matter whether angels or devils uncover how the world works. It doesn't even matter if they give proper credit to others . . .

ULLA ZORN

You're pretty cynical.

BENGT HJALMARSSON

I've heard that before . . . from another woman. But another part of me also knows that science is not always about power . . . or control . . . or even progress. The world can be a playground, full of secrets. Where it's just plain fun for me to find out why one molecule is bent, another linear.

(Hjalmarsson imitates the bending of a molecule with his hands)
(Astrid Rosenqvist enters, but at first is not noticed by Zorn and Hjalmarsson).

ULLA ZORN

(Slightly ironic, yet touched)
In other words, scientists can also have fun . . . like historians?

BENGT HJALMARSSON

(Startled)
Astrid! What brings you back?

ASTRID ROSENQVIST

(Sensing both their discomfort at having been overheard)
I forgot my cigarettes.
(Points to cigarette case lying on table near her seat)

BENGT HJALMARSSON

Well . . . so long to both of you . . . I've got to get back to the lab.
(Exits)

ASTRID ROSENQVIST

Well, Ulla . . . what do you think now?

ULLA ZORN

(Flustered)
What do you mean?

ASTRID ROSENQVIST

Of Bengt.

ULLA ZORN

As you said earlier . . . an interesting man.

ASTRID ROSENQVIST

I think I said "complicated."
(Looks pensively at Ulla Zorn)
But I agree . . . he's also interesting . . . even now.

ULLA ZORN

Compared to when?

ASTRID ROSENQVIST

I'll tell you a story about him. You know he was at the Pasteur Institute. He met there a young French biologist, brought her back to Sweden.

SCENE 8

ULLA ZORN

(Intensely curious)

Now that is interesting!

ASTRID ROSENQVIST

They lived together, but she couldn't take our November nights. So she went back to France. Since then he's been in the lab all the time. But at night he plays the cello.

ULLA ZORN

So you like Bengt?

ASTRID ROSENQVIST

You do too . . . don't you?

LIGHTS FADE

END OF SCENE 8

1777

SCENE 9

(Stockholm, 1777, evening following Stockholm Challenge in Scene 8). Bare room, very dark, three couples (left upstage, center downstage, and right upstage) barely discernible.

(Spotlight focuses on MRS. PRIESTLEY *and* PRIESTLEY. *They whisper)*

MRS. PRIESTLEY
Why face him?

PRIESTLEY
I must.

MRS. PRIESTLEY
To prove you told him?

PRIESTLEY
To show I was first.

MRS. PRIESTLEY
And Scheele?

PRIESTLEY
I trust him.

MRS. PRIESTLEY
He claims priority.

PRIESTLEY
He did not publish.

MRS. PRIESTLEY
Yet wasn't he first?

PRIESTLEY
Perhaps.

MRS. PRIESTLEY
But that would make you second.

PRIESTLEY
It would make Lavoisier third.

MRS. PRIESTLEY
Is that the point? That he was last.

PRIESTLEY
Indeed.

MRS. PRIESTLEY

Why?

PRIESTLEY

Have the world bow to him? *(Pause)*. When I preceded him?

MRS. PRIESTLEY

If you were King Gustav –

PRIESTLEY

God forbid!

MRS. PRIESTLEY

(Persists)

Still . . . if you were King . . . whom would you pick?

PRIESTLEY

I'd ask . . . whom would the world choose?

MRS. PRIESTLEY

Joseph! Answer me . . . as my husband . . . not as a clever minister.

PRIESTLEY

You've always wanted black and white answers.

MRS. PRIESTLEY

This question deserves it.

PRIESTLEY

Deserving something does not always lead to getting it.

MRS. PRIESTLEY

You're not in a pulpit.

PRIESTLEY

(Tired)

I published first . . . which makes me first in the world's eyes.

MRS. PRIESTLEY

I meant the heart . . . not the eyes.

PRIESTLEY

The world has no heart.

MRS. PRIESTLEY

But you do . . . you've often opened it to me.

PRIESTLEY

You're a clever woman, Mary.

MRS. PRIESTLEY

No . . . this is your loving wife asking.

PRIESTLEY

Before we came to Stockholm, I was convinced . . . in my heart and mind . . . that I was first. *(Pause)*. But now?

MRS. PRIESTLEY

I understand, Joseph.

Spotlight focuses on FRU POHL *and* SCHEELE. *They whisper.*

FRU POHL

How will you convince him?

SCHEELE

I have a copy.

FRU POHL

He may claim you never sent it.

SCHEELE

Bergman saw it.

FRU POHL

I forgot.

SCHEELE

He will remember.

FRU POHL

There is Madame Lavoisier–

SCHEELE

Yes?

FRU POHL

I think she knows.

SCHEELE

You are sure?

FRU POHL

I sensed it.

SCHEELE

Not much is hidden in a sauna.

FRU POHL

But did she tell her husband?

SCHEELE

Would you have?

FRU POHL

(Hesitates)

Yes.

SCHEELE

Why?

FRU POHL

Because . . . it would be right.

SCHEELE

You're a good woman . . . Sara Margaretha.

FRU POHL

You told me that once before.

SCHEELE

It bears repeating.

FRU POHL

Mme. Lavoisier . . .

SCHEELE

I do not trust her.

FRU POHL

But does he?

(Spotlight focuses on MME. LAVOISIER *and* LAVOISIER. *They whisper.)*

MME. LAVOISIER

You will meet them both?

LAVOISIER

His Majesty insisted.

MME. LAVOISIER

But that dinner in Paris with Priestley . . . I worry.

LAVOISIER

So do I. There were witnesses.

MME. LAVOISIER

And the letter?

LAVOISIER

What letter?

MME. LAVOISIER

Scheele's. I saw it . . .

LAVOISIER

(Taken aback)

You saw it?

MME. LAVOISIER

But I couldn't tell you.

LAVOISIER

(Furious)

Why do it now?

MME. LAVOISIER

I feel guilty.

LAVOISIER

(Even angrier)

And I must share your guilt?

MME. LAVOISIER

You are my husband.

LAVOISIER

Where is the letter?

MME. LAVOISIER

Hidden.

LAVOISIER

(Taken aback)

You have not destroyed it?

MME. LAVOISIER

You look so angry. Why?

LAVOISIER

I do not wish to discuss it.

MME. LAVOISIER

You cannot tell your wife?

LAVOISIER

I can tell no one.

MME. LAVOISIER

But why?

LAVOISIER

Once voiced, I'd have to deny that thought ... or condemn it.

MME. LAVOISIER

So you disapprove of what I did?

LAVOISIER

You're still young.

MME. LAVOISIER

Why blame youth?

LAVOISIER

Subtlety only comes with maturity.

MME. LAVOISIER

You taught me chemistry ... now teach me subtlety.

LAVOISIER

Subtlety cannot be taught.

MME. LAVOISIER

Nor explained?

LAVOISIER

If I'd known he'd choose a personal letter – not a scientific article – to establish his priority, I would have wished that letter away.

MME. LAVOISIER

Of course! That's why–

LAVOISIER

Wait! But not wished to know how it disappeared.

MME. LAVOISIER

If that is subtlety ... I do not understand it.

LAVOISIER

A stray thought becomes iniquity when spoken.

MME. LAVOISIER

It's the lawyer in you ... a role I never liked.

LAVOISIER

The law is never likeable ... especially when dealing with guilt.

MME. LAVOISIER

I'm the guilty one ... I admitted it ... and only to you.

LAVOISIER

Tainted by knowledge of the deed, how can I approve of my wife's action?

MME. LAVOISIER

Even when done as a token of love ... for you?

LAVOISIER

Especially if done for love ... for then I must reject your gift of love as well.

BLACKOUT. *(Women exit)*

LIGHT on SCHEELE, LAVOISIER, and PRIESTLEY

SCHEELE

"Resolve the question: Who made fire air first?" That was His Majesty's command.

LAVOISIER

Is that the real question?

PRIESTLEY

Of course. And you, Monsieur ... did not make that air first ... as you yourself in effect conceded yesterday.

LAVOISIER

I understood it first ...

SCHEELE

Understanding only comes after existence!

PRIESTLEY

But proof of such existence must be shared!

SCHEELE

I shared it! Here is the letter ...

(Offers it to Lavoisier who does not take it)

dispatched to you almost three years ago. Describing work done earlier still.

LAVOISIER

(Aggressive, yet words carefully chosen)

I never heard about that letter until today ...

SCHEELE

It describes the preparation of fire air ...

LAVOISIER

No such letter ever reached me.

SCHEELE

A recipe you reproduced in front of all of us today.

LAVOISIER

But certainly not years ago as you now claim.

(Impatient)

What is the real purpose of this meeting?

PRIESTLEY

Priority! In August 1774 I made dephlogisticated air . . . your oxygen . . .

LAVOISIER

Then you thought you had nitrous air, sir.

PRIESTLEY

The first steps of discovery are often tentative.

LAVOISIER

Some of us are more careful than others.

PRIESTLEY

In October of that year, I met the leading chemists of France *(Pause)*. . . including you, Sir.

LAVOISIER

Indeed, you dined in my house in Paris.

PRIESTLEY

I told the gathering . . .

LAVOISIER

in your imperfect French . . .

PRIESTLEY

. . . which Madame Lavoisier comprehended fully . . . of my discovery.

LAVOISIER

Your report . . . lacked clarity. Your methods were imprecise . . .

PRIESTLEY

Sir, your words are unworthy.

LAVOISIER

At best, you, Dr. Priestley, supplied us with the smallest of clues . . .

PRIESTLEY

I thought details mattered to you, sir.

LAVOISIER

Only if they are relevant.

PRIESTLEY

More than once, my experiments in pneumatic chemistry were cited by you–

LAVOISIER

Is that a reason to complain?

PRIESTLEY

Only to be then diluted ... if not evaporated.

LAVOISIER

How did I do so?

PRIESTLEY

You write

(heavy sarcasm)

"We did this ... and we found that." Your royal "we", sir, makes my contributions disappear ... poof ... into thin air! *(Pause)*. When I publish, I say, "I did ... I found ... I observed." I do not hide behind a "we."

LAVOISIER

Enough of generalities ... *(Aside)*... or platitudes. *(Louder)*. What now?

PRIESTLEY

The question, sir! The question! Who made that air first?

SCHEELE

(Much more insistent than before, to audience)

I did. And future generations will affirm it.

PRIESTLEY

(To audience)

But by the grace of God, I made it too ... and published first!

LAVOISIER

(To audience)

They knew not what they'd done . . . where oxygen would lead us.

(The three men start arguing simultaneously in loud tones so that the words cannot be understood)

OFFSTAGE

COURT HERALD'S VOICE

(Tapping of staff)

Three savants? Yet you cannot agree? So be it. *(Pause)*. The king will not reward you!

END OF SCENE 9

INTERMEZZO 4

Immediately following Scene 9
(Stockholm, 1777, sauna)

FRU POHL
So we meet once more before your departure.

MRS. PRIESTLEY
(Playing with birch branch in hand)
Perhaps the last sauna of my life.

FRU POHL
Madame Lavoisier declined. Perhaps today she has something to hide.

MRS. PRIESTLEY
The Judgment of Stockholm didn't please her. Yet who cares if the King decides . . . or doesn't?

FRU POHL
Herr Scheele and I care. He is our King.

MRS. PRIESTLEY
The rewards of discovery are not of this earth.

FRU POHL
The words of a minister's wife. But Apothecary Scheele seeks the affirmation of his peers.

MRS. PRIESTLEY
He has their respect.

FRU POHL
In his quiet way, he wishes for more. And . . . we need new shelves for our pharmacy . . . The King's reward was not solely a medal.

MRS. PRIESTLEY
Surely your friends . . .

FRU POHL
Will they? *(She shifts discussion)*. Our men, who actually made the fire air . . .

MRS. PRIESTLEY

(Lightly strikes Fru Pohl with the birch branch)

... dephlogisticated air, if you will.

FRU POHL

You see? We are just like them. Call it what you may...

MRS. PRIESTLEY

Even the Frenchman's "oxygen"?

FRU POHL

Yes, even that. What matters is ... they could not agree on who made it first.

MRS. PRIESTLEY

Will they ever?

FRU POHL

I doubt it. They missed their chance.

MRS. PRIESTLEY

So it will remain a mystery?

FRU POHL

Oh no. The world likes it simple. Others will decide.

END OF INTERMEZZO

SCENE 10

(Stockholm, 2001; Royal Academy of Sciences, two weeks after Scene 8. Ulla Zorn is fiddling with some projection equipment for a computer-controlled presentation, which none of the others notice).

ASTRID ROSENQVIST

It's time for a formal motion. *(Looks around)*. Bengt ... do you want to start?

BENGT HJALMARSSON

(With exaggerated formality)

I propose that the Royal Swedish Academy of Sciences select Antoine Laurent Lavoisier, the architect of the Chemical Revolution, for the first retro-Nobel Prize in Chemistry.

(Reverts to ordinary tone).

I hope that's formal enough.

SUNE KALLSTENIUS

Mine is less formal, but direct: I propose Carl Wilhelm Scheele for first discovering oxygen. *(Pause)*. A humble man too, who didn't indulge in hype or self-promotion.

ULF SVANHOLM

Clearly, we have a case of simultaneous discovery. What are a few months between friendly competitors? So let me get right to the point: I propose Scheele and Priestley. Period! But Lavoisier? He may deserve it – but not for the discovery of oxygen.

BENGT HJALMARSSON

I nominated him as the Father of the Chemical Revolution ... which happened to have come from oxygen! Lavoisier's moral failures are clear, yes ... but he brought about true change, by making chemists pay attention to the balance sheet of nature!

ULF SVANHOLM

Ignore moral lapses?

BENGT HJALMARSSON

It's happened more than once with our regular Nobel Prizes. Good or poor ethics simply can't be weighed on the same scale with good or poor science!

ULF SVANHOLM

But what a precedent for the first retro-Nobel!

ASTRID ROSENQVIST

Please! Today, we deal with nominations . . . not with reasons for the nominations.

BENGT HJALMARSSON

And whom do you favor, oh chair? So far, you've acted like a sphinx.

ASTRID ROSENQVIST

There are seven combinations of the three names that one might imagine: Three men alone . . . three pairs . . . and all three together. Let me put on the table the most attractive option . . . awarding the first retro-Nobel Prize in Chemistry jointly to all three. But cite them for the Chemical Revolution rather than the discovery of oxygen.

ULF SVANHOLM

Include Lavoisier . . . who failed to credit work explicitly reported to him by Priestley . . . and confided to him in Scheele's letter?

ULLA ZORN

Which Lavoisier had never seen.

SUNE KALLSTENIUS

What? Ms. Zorn . . . what did you say?

ULLA ZORN

Your work intrigued me . . . for it continually crossed the lives of women in my thesis. So I made a quick flight to Cornell University in Ithaca, New York.

BENGT HJALMARSSON

I know all about the Cornell collection of Lavoisier papers. What could you have found there?

ULLA ZORN

(Quietly, yet triumphantly)

A book.

BENGT HJALMARSSON

(Sarcastic)

What a surprise! Finding a book . . . in a library.

ULLA ZORN

A book entitled "Histoire des Théatre."

SUNE KALLSTENIUS

How could such a book help us?

ULLA ZORN

Let me show you some slides.

(She presses a couple of keys on her computer)

(Fig. 3a)

The object I found there only looks like a book.

(Fig. 3a appears on the screen, the first picture of the nécessaire, closed, in the hands of a woman, causing general

consternation among the committee members except Astrid, who is smiling)

It's Madame Lavoisier's *nécessaire* . . . a travel chest, disguised as a book. To my knowledge, historians had not mentioned it before. But I saw it in the catalog of the 1956 "Souvenir de Lavoisier" sale held in Paris. And then found out that the Cornell Library had purchased it in 1963. *(Pause)*. So I decided to take a look–

ULF SVANHOLM

Intuition?

ULLA ZORN

(Sharply)

Why not call it hands-on research by a historian? Here it is opened

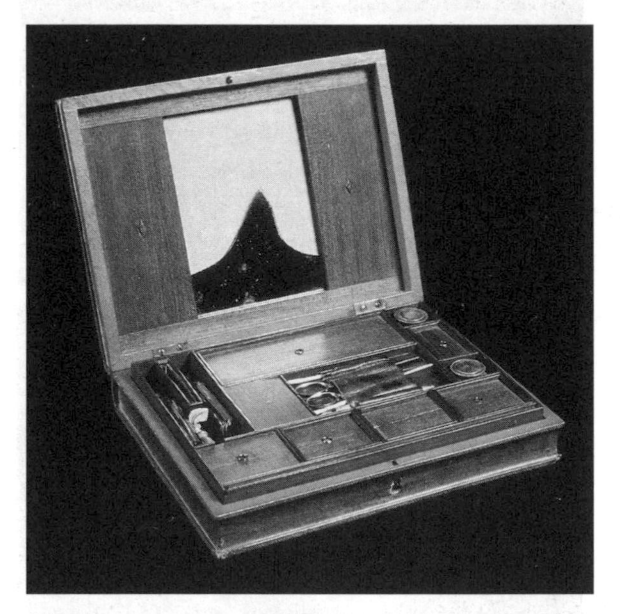

(Fig. 3b)

Look at all the compartments,

(Uses laser pointer to identify the various items)

. . . with thread, needles, combs, pens, and bottles for perfumes and ink . . .

(Fig. 3c)

Even a ruler, crammed in a slit, like in a Swiss Army knife.

ULF SVANHOLM

I'll be damned!

ULLA ZORN

(Pleased and excited)

When you lift the tray out,

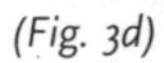

(Fig. 3d)

there is place for stationery. I checked the watermarks. The paper is actually after Mme. Lavoisier's times ... her heirs must have used the *nécessaire*. The broken mirror in the lid of the box intrigued me ...

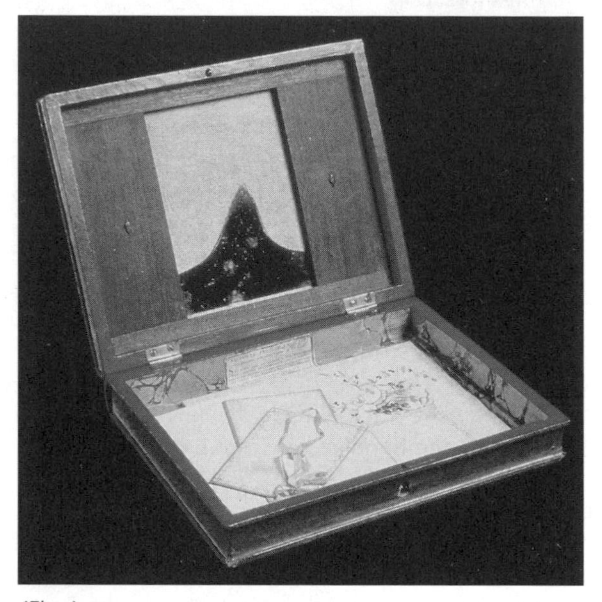

(Fig 4)

There was a space behind the mirror. We poked gently around it; the curator as excited as I. We found a paper. This one ...

(Waves paper in the air)

SUNE KALLSTENIUS

Stop teasing us! What is it?

ULLA ZORN

A letter ... this is a photocopy, of course ... a letter apparently never sent. *(Pause)*. From Madame Lavoisier ... to her husband.

ULF SVANHOLM

And how do you know it's Mme. Lavoisier's?

ULLA ZORN

I had an expert at Cornell check it.

BENGT HJALMARSSON

(Impatiently)

What did it say?

ULLA ZORN

She writes . . .

(At this point the LIGHTS DIM *on the frozen committee, except for Ulla Zorn.* LIGHTS RISE *on Mme. Lavoisier, upstage)*

MME. LAVOISIER

My dear husband. In these difficult times, in the separation forced upon us by the Revolution, I reflect on the past. I return time and time again to Apothecary Scheele's 1774 letter . . .

ULLA ZORN

Apparently she intercepted Scheele's famous letter . . . remember she handled much of Lavoisier's correspondence.

MME. LAVOISIER

Now that the brilliance and accuracy of your studies have convinced the world of the central role of oxygen in chemistry, now that phlogiston lies in the dustbin of discarded theories . . . I will not speak of the diehards such as Dr. Priestley who continue to preach it. *(Pause)*. I ask you now to forgive me. I could not show Apothecary Scheele's letter to you, my dear husband. It would have taken the wind out of your sails, you, who were so close . . . And I told you why I felt incapable of destroying it. Our priority rested on my hiding it.

*(*LIGHTS FADE *on Mme. Lavoisier)*

ULLA ZORN

Note! She didn't say "your priority" . . . but "our." She filed the letter, without showing it to him. Or actually misfiled it, which may be one of the reasons why it surfaced only over 100 years later when Grimaux found it

BENGT HJALMARSSON

And you waited until now to tell us?

ASTRID ROSENQVIST

She told me –

BENGT HJALMARSSON

(Outraged)

And why not me . . . or the rest of us?

ASTRID ROSENQVIST

I thought Ulla was entitled to announce her discovery herself. If anyone is to blame . . . blame me.

BENGT HJALMARSSON

The question was not addressed to you!

(Faces Ulla Zorn)

Why? To show how clever you are? *(Softer)*. I would have told you that . . . on my own . . . if you'd had the courtesy to inform me first . . . Wasn't Lavoisier my responsibility?

ULLA ZORN

I didn't mean to trespass on your turf.

BENGT HJALMARSSON

That's water over the dam.

ULLA ZORN

I meant to help . . .

BENGT HJALMARSSON

(Warmer tone)

But why was her letter in that *nécessaire*? Why was it never sent?

ULLA ZORN

I asked myself that same question.

BENGT HJALMARSSON

And?

ULLA ZORN

I didn't tell you yet the date of Mme. Lavoisier's letter. It was just before Christmas, 1793, when Lavoisier is in jail, a few months before his execution.

BENGT HJALMARSSON

(Gently)

Nineteen years after Scheele's letter came.

ULLA ZORN

Evidently, she continued to brood about it. During the worst of times, with her husband in prison . . . she wrote to him . . . returning to what she had done years earlier. But when she did, it was no longer safe to send that letter.

(Ulla Zorn sits back. The committee is pensive).

BENGT HJALMARSSON

One letter she could not send . . . another she could not burn.

END OF SCENE 10

INTERMEZZO 5

(After scene 10)

Funereal darkness except for spotlight on Mme. LAVOISIER *on extreme left downstage, quill pen in hand about to write a letter and M.* LAVOISIER *extreme right downstage. Each in apparent soliloquy)*

MME. LAVOISIER

My husband ... you recognized a girl's talent ... And like my father ... you did not snuff it out.

LAVOISIER

My dear wife ... In the solitude of a cell ... I think of our life together.

MME. LAVOISIER

You were not bored when I played the harp for you in my father's house ...

LAVOISIER

You were not bored when I spoke about geology ... about chemistry ...

MME. LAVOISIER

When we played the "Game of Good Fortune"... I wondered where the arrow would stop. At which word? "Wisdom" ...? "Convent" ...? "Marriage" ...?

LAVOISIER

I hid from you a magnet's power ... whereby I moved the arrow to *(Pause)* ...

MME. LAVOISIER

Or perhaps ...

LAVOISIER and MME. LAVOISIER

(In unison)

"Love" ...

LAVOISIER

... a word I had not used before. And then I married you ...

MME. LAVOISIER

But ...

LAVOISIER

... to become my trusted partner ...

MME. LAVOISIER

... I never heard you say "love" again.

LAVOISIER

I had no time for idle pastimes ... even for children. I thought you understood ...

MME. LAVOISIER

Science and public service were your *métier.* *(Pause)*. Still ...

LAVOISIER

I always felt you were satisfied, yet for you ...

MME. LAVOISIER

... something was missing.

LAVOISIER

... there were other men.

MME. LAVOISIER

The word your arrow showed me in my father's house ...

LAVOISIER

Love? *(Pause)*.

MME. LAVOISIER

... is what I missed.

LAVOISIER

No, I offered you more. True partnership. *(Pause)*. No other man could do the same ...

MME. LAVOISIER

Pierre Du Pont offered love ... for seventeen years. No matter ... *(Pause)* I did not dare explain to you ...

LAVOISIER

In prison I now understand ...

MME. LAVOISIER

... what I had done.

LAVOISIER

... what I had neglected ...

MME. LAVOISIER

(Reaches for a sheet of paper)

Now I must write of it.

LAVOISIER

. . . to appreciate:

MME. LAVOISIER

Before it is too late.

LAVOISIER

Ambition without love is cold.

MME. LAVOISIER

I've never loved another man.

BLACKOUT

SCENE 11

(Stockholm, 2001, Committee Room. SVANHOLM *sits morosely at table.* KALLSTENIUS *enters).*

SUNE KALLSTENIUS

That was a pretty good paper you published the other day. The one on polycarbonates.

ULF SVANHOLM

(Suspicious)

Pretty good?

SUNE KALLSTENIUS

All right . . . damned good.

ULF SVANHOLM

That's better. But why the compliment?

SUNE KALLSTENIUS

It wasn't meant as a compliment . . . it's a factual observation.

ULF SVANHOLM

(Pleased)

You mean that? *(Brief pause)*. But why tell me now?

SUNE KALLSTENIUS

Astrid was right . . . "bury the hatchet."

ULF SVANHOLM

Hmm.

SUNE KALLSTENIUS

Just "hmm"? Ulf . . . you're carrying a grudge too far.

ULF SVANHOLM

I?

SUNE KALLSTENIUS

Okay . . . okay. We.

ULF SVANHOLM

That's better.

SUNE KALLSTENIUS

You've always blamed me for holding up your paper.

ULF SVANHOLM

You did! For 6 months!

SUNE KALLSTENIUS

Don't start that all over again.

ULF SVANHOLM

That too was first-class research!

SUNE KALLSTENIUS

My job as a reviewer was to question the evidence.
Even of first-class research ...

ULF SVANHOLM

Your so-called "improvements" enabled your pals at Stanford to scoop us.

SUNE KALLSTENIUS

I knew nothing about the Stanford work.

(Conciliatory)

Ulf, I didn't talk to them. You can't go on blaming me.

(SVANHOLM, with sullen expression, rises and starts to pace up and down. Suddenly he notices the masks on the wall)

ULF SVANHOLM

(Lifts a mask off the wall)

Look at this! Did you see these before?

SUNE KALLSTENIUS

A mask. They've been there all the time.

ULF SVANHOLM

(Lifts up another)

Two masks.

SUNE KALLSTENIUS

Let's try them on.

ULF SVANHOLM

It seems silly.

SUNE KALLSTENIUS

Who will see us?

ULF SVANHOLM

In this Committee ... in science ...

SUNE KALLSTENIUS

Yes?

ULF SVANHOLM

Most wear masks anyway ...

SUNE KALLSTENIUS

The scientists' "Masked Ball?"

ULF SVANHOLM

Exactly! With masks on, we can pretend we're impartial.

SUNE KALLSTENIUS

Rational gentlemen scholars, eh?

ULF SVANHOLM

Yes. But I'm not that way.

SUNE KALLSTENIUS

(Bantering)

You're not rational . . . or you're not a gentleman? *(Laughs)*. Or is it both?

ULF SVANHOLM

Wearing a mask, I'd probably concede that those guys at Stanford did that research independently. But without it, I'm screaming: "They stole my catalyst!" And *(grimace)* that you helped them.

SUNE KALLSTENIUS

My catalyst? You sound like one of them.

ULF SVANHOLM

Them?

SUNE KALLSTENIUS

My fire air . . . my dephlogisticated air . . . my oxygen . . .

ULF SVANHOLM

(Plays with the masks)

In that case, let's put them on. Let's pretend we've just discovered oxygen.

SUNE KALLSTENIUS

I'll take the one with the sun and fire.

ULF SVANHOLM

And leave me with rust and ashes?

(They slip on the masks and pretend they're fencing. Astrid Rosenqvist enters).

ASTRID ROSENQVIST

What on earth are you doing?

(Both men, embarrassed, pull off their masks).

ULF SVANHOLM

Playing.

ASTRID ROSENQVIST

Live and learn! I thought you were fighting . . . as usual.

SUNE KALLSTENIUS

He came to his senses.

ULF SVANHOLM

And he, for a change, liked some work I did.

ASTRID ROSENQVIST

If you two really made up, I don't need to know why. Just do me a favor, both of you: agree on one candidate for the retro-Nobel. It will simplify life for me.

SUNE KALLSTENIUS

You aren't trying to lead us down a garden path by any chance . . . are you?

ASTRID ROSENQVIST

Me . . . an innocent theoretical chemist?

SUNE KALLSTENIUS

Yes . . . you. You're pushing for a consensus, when we should be making a tough choice: one winner only. Take the Literature Nobel. It's never shared!

ASTRID ROSENQVIST

But that's preposterous! It's comparing watermelons with . . . *(searches for the right word, finally finds it)*. . . peanuts!

ULF SVANHOLM

I suppose literature is peanuts.

SUNE KALLSTENIUS

(Infuriated)

I'm dead serious.

ULF SVANHOLM

But so am I. You're ignoring two fundamental differences between literature and science. The literati don't worry about priority . . . and if they'd had a retro in Literature, it would've gone to Shakespeare or Dante or Cervantes . . . or whoever . . . but it wouldn't be shared. If Shakespeare had never lived, "King Lear" could never have been written. Without Dante, there would be no "Divine Comedy." Without Cervantes –

SUNE KALLSTENIUS

Ulf, what's your point?

ULF SVANHOLM

Simple! Consider oxygen. If Scheele or Priestley or Lavoisier had never lived, somebody would have discovered oxygen. The same with Newton and gravity, with Mendel and genetics–

SUNE KALLSTENIUS

So why give a Nobel at all in your water melon patch? If it would happen anyway, why worry who is first?

ULF SVANHOLM

Because science is done by scientists . . . not machines . . . and scientists crave recognition.

ASTRID ROSENQVIST

Science is done by humans . . . humans are competitive . . . scientists are even more competitive . . . and they want to be rewarded for being first.

SUNE KALLSTENIUS

Of course! But we still haven't agreed what "being first" means: is it the initial discovery . . . the first publication . . . or full understanding?

(Bengt Hjalmarsson and Ulla Zorn wander in during the preceding exchange, but are not noticed by the others until they speak)

BENGT HJALMARSSON

(Ironic)

Let's see, where was it that Columbus thought he was sailing?

ULF SVANHOLM

(Snaps at Hjalmarsson)

Who cares? Our Vikings got there first . . .

ULLA ZORN

To find people who had come thousands of years earlier . . .

END OF SCENE 11

SCENE 12

(Stockholm, 2001. Royal Academy of Sciences, a few minutes after scene 11. ASTRID ROSENQVIST *sits on edge of table, one leg exposed to lower thigh).*

BENGT HJALMARSSON

(Points to her slit skirt)

You're still a damned good-looking woman, Astrid . . .

ASTRID ROSENQVIST

I wore the skirt for you.

BENGT HJALMARSSON

Well . . . It worked . . . It got a rise out of me.

ASTRID ROSENQVIST

That's exactly what you said the first time we met . . . way back when.

BENGT HJALMARSSON

The chemistry was right . . . then.

ASTRID ROSENQVIST

Then . . .

(Awkward pause for both).

Like it was for the Lavoisiers, when they were young . . .

BENGT HJALMARSSON

But how did they manage not to have any children?

ASTRID ROSENQVIST

(Laughs)

Not the way we did!

BENGT HJALMARSSON

So what was their secret?

ASTRID ROSENQVIST

Maybe one of them was infertile.

BENGT HJALMARSSON

Or . . . *(does not finish the sentence)*

ASTRID ROSENQVIST

Go on . . . finish the sentence.

BENGT HJALMARSSON

Maybe the marriage was never consummated . . .

ASTRID ROSENQVIST

But you just read us a pretty passionate love letter.

BENGT HJALMARSSON

That was from Du Pont.

ASTRID ROSENQVIST

Maybe he also played the cello . . . Cello and physical chemistry . . . an irresistible combination.

BENGT HJALMARSSON

The cello was for pleasure . . . yours and mine.
Something else got in our way . . .

ASTRID ROSENQVIST

An ambitious man always has problems–

BENGT HJALMARSSON

With an ambitious woman.

ASTRID ROSENQVIST

What else is new? *(Pause)*. We're smart. *(She speeds up)*.
We even want to do something
(Draws quotation marks in air)
"for the benefit of mankind."

BENGT HJALMARSSON

And we want the world to know it.

ASTRID ROSENQVIST

Yes. Somehow I thought the retro-Nobel for the dead would be . . . purer.

BENGT HJALMARSSON

You were wrong.

ASTRID ROSENQVIST

At least the retro-Nobel got us into the same room . . .

BENGT HJALMARSSON

As chair . . . you could've asked for someone else.

ASTRID ROSENQVIST

You could've refused. Why didn't you?

BENGT HJALMARSSON

For the same reason that you didn't pick a substitute.

ASTRID ROSENQVIST

Then . . . why are you so prickly during our meetings?

BENGT HJALMARSSON

And why are you so bossy?

ASTRID ROSENQVIST

We should learn to compromise . . .

BENGT HJALMARSSON

Something neither one of us was good at. Is it possible to change?

(Begins to exit, but as he passes her, he displays some paternal, non-erotic gesture–perhaps kissing her on her head and exits).

Astrid Rosenqvist slowly walks to the table. She pulls out a pack of cigarettes, looks at it, but decides against smoking and tosses it on the table near her chair. ULF SVANHOLM joins her.

ULF SVANHOLM

(Points to her cigarette pack)

Are you kicking the habit?

ASTRID ROSENQVIST

Not yet. But Ulla's discovery feels like nicotine. I'm relieved that Lavoisier never saw Scheele's letter.

ULF SVANHOLM

Does it change the facts? We all know Lavoisier was not the first to discover oxygen!

ASTRID ROSENQVIST

You still have to understand what you discover. Do you realize that as late as 1800 your man Priestley still wrote a book entitled "The Doctrine of Phlogiston Established and that of the Composition of Water Refuted"? *(Pause)*. In other words, "down with H_2O" but "onwards with mumbo jumbo."

ULF SVANHOLM

You're too hard on my experimentalist.

ASTRID ROSENQVIST

The world needs physical chemists, like Lavoisier . . . or even better . . . theoreticians.

ULF SVANHOLM

Like you?

ASTRID ROSENQVIST

They could have done worse . . . but we all know what role women played in chemistry at that time. Madame Lavoisier got about as close as was realistic.

LIGHTS FADE,
allowing committee to reassemble
LIGHTS UP

ASTRID ROSENQVIST

Now let's all sit down and arrive at a decision.

(All head for the table except for HJALMARSSON, *who walks toward* ULLA ZORN, *waiting until she looks up at him from her computer).*

BENGT HJALMARSSON

(Low voice)

I owe you an apology about Mme. Lavoisier's letter. I was boorish . . .

ULLA ZORN

(Pleased)

I would have used another word . . . but . . . *(pause)* thanks . . .

BENGT HJALMARSSON

May I pay you a compliment?

ULLA ZORN

(Playfully)

Do you think I'll be able to handle it?

BENGT HJALMARSSON

(Seriously)

I wish I'd found that travel chest . . .

ULLA ZORN

(Pleased)

That is a compliment!

BENGT HJALMARSSON

Ulla. *(Hesitates, drops voice)*. May I invite you –

ASTRID ROSENQVIST

(Who has overheard them as she approaches table, sharply)

Bengt! First things first! Would you please join us?

(Points to conference table)

BENGT HJALMARSSON

(Touch of irony)

"First" in the chair's eyes or mine?

(Rosenqvist waits until Hjalmarsson sits down)

ASTRID ROSENQVIST

We have four committee members . . . and four proposals: Lavoisier alone . . . Scheele alone . . . Priestley plus Scheele . . . and finally all three together. I presume that each of you would still vote for his original recommendation?

(Everyone nods assent).

That won't get us very far. We've got to come up with a consensus for the academy–

BENGT HJALMARSSON

Or at least a majority.

ASTRID ROSENQVIST

A consensus would be far preferable . . . at least for the first retro-Nobel.

(Looks around)

In that case let's vote by formal written ballot–

ULF SVANHOLM

We've never before had written ballots in Nobel committees.

ASTRID ROSENQVIST

For a retro-Nobel committee there is no precedent.

BENGT HJALMARSSON

You want us to vote for our second choices? What if there aren't any?

ASTRID ROSENQVIST

(Sharply)

You . . . more than anyone else in this room . . . should know that in life, we mostly end up with second choices.

BENGT HJALMARSSON

(Mockingly miming her tone)

You . . . more than anyone else in this room . . . should know that I can't be forced into a decision.

ASTRID ROSENQVIST

(Equally mocking tone)

Which will never stop me from trying to persuade you . . . all of you . . . to arrive at a consensus. Otherwise, all we can forward to the Academy is a recommendation that the discovery of oxygen be honored with the first retro-Nobel but leaving it to the Academy members to vote who of the three candidates should get it.

ULF SVANHOLM

I'd rather do that than compromise.

ASTRID ROSENQVIST

I'd rather not . . . it would be embarrassing.

ULF SVANHOLM

For the chair?

ASTRID ROSENQVIST

For all of us. *(Pause)*. Listen! Here is a way to resolve our problem: We all vote for a pair of candidates.

SUNE KALLSTENIUS

(Puzzled)

How will voting for pairs help?

ULF SVANHOLM

In other words, only three options? Lavoisier-Scheele, Lavoisier-Priestley and Priestley-Scheele. But no single choices?

ASTRID ROSENQVIST

Precisely!

BENGT HJALMARSSON

(Disdainfully)

Brilliant! But why go through that exercise?

ASTRID ROSENQVIST

My original proposal . . . anointing all three . . . would,

of course, be simplest. But since none of you seems to go for that, voting for pairs at least forces everyone to think about another candidate . . . while still holding on to their favorite.

*(*KALLSTENIUS *and* SVANHOLM *look at* ROSENQVIST. *One shrugs, the other nods. Long pause)*

ASTRID ROSENQVIST

Bengt?

*(*HJALMARSSON *looks at her but says nothing, whereupon* ROSENQVIST *rises and walks to him. Continues in low voice)*

We both know what Lavoisier did.

BENGT HJALMARSSON

So?

ASTRID ROSENQVIST

Do we really dilute Lavoisier's credit by adding one other man? Earlier on, you said neither one of us was good at compromise. How about demonstrating that you were wrong?

*(*HJALMARSSON *shrugs, followed by reluctant nod, then turns to look away.* SVANHOLM *approaches* KALLSTENIUS*)*

ULF SVANHOLM

(Whispers)

Did you hear what she said?

SUNE KALLSTENIUS

I sure did.

ULF SVANHOLM

Out of the question! If Lavoisier gets the nod, then only Scheele or Priestley can share it.

SUNE KALLSTENIUS

I could live with that . . . if Scheele is the other.

ULF SVANHOLM

But what if he is not? What if I vote for Lavoisier and Priestley and so do they?

SUNE KALLSTENIUS

I'd object!

ULF SVANHOLM

A lot of good that will do you . . . after the vote has been counted.

SUNE KALLSTENIUS

So what's your proposal?

ULF SVANHOLM

Let's both vote for your man . . . and my Priestley.

SUNE KALLSTENIUS

Hmm.

ULF SVANHOLM

What does that mean?

SUNE KALLSTENIUS

You'll see. Wait 'till we've voted.

ASTRID ROSENQVIST

(To Zorn)

Ulla . . . would you distribute the ballots?

(After having given ballots to KALLSTENIUS *and* SVANHOLM, ZORN *heads for* HJALMARSSON, *but* ROSENQVIST *intercepts her. Takes ballot and hands it personally to* HJALMARSSON*)*

(Very gently).

Please, Bengt . . . please. Two names. Do Madame Lavoisier that favor.

(Hjalmarsson looks at her, then takes the paper, but freezes as the shadowy figure of MME. LAVOISIER *appears.*
As she speaks, she approaches Hjalmarsson until she practically touches him)

FAINT LIGHT on MME. LAVOISIER

MME. LAVOISIER

(Addresses Hjalmarsson, who does not see her)

Perhaps King Gustav will reconsider. Important decisions shouldn't be left to others. Least by a king.

(Pause).

(Shakes her head, firmer)

It won't matter... Posterity will recognize my husband. *(Pause)*. Of course... some will ask: What good is such recognition?
(Smiles to herself)
Much good will come from our oxygen... kings will surely tax it.
(Pause, then turns serious)
But after death? Our children *(she shivers)*... will carry on where the gauche apothecary... and the priestly chemist... and my husband stopped. *(Pause)*. Imagine what it means to understand what gives a leaf its color! What makes a flame burn. Imagine!

END OF SCENE 12

END OF PLAY

ACKNOWLEDGMENTS

Thanks are due to to Lavinia Greenlaw for rendering difficult concepts into verse in Scene 6; to the Ithaca College Department of Theatre Arts and the Kitchen Theatre for organizing an early reading of OXYGEN in Ithaca, NY; to PlayBrokers for a staged rehearsed reading (directed by Ed Hastings) at the ODC Theatre in San Francisco; to Nicholas Kent for arranging a staged rehearsed reading (directed by Erica Whyman) at the Tricycle Theatre in London; and to the Department of Theatre, Film & Dance at Cornell University for a reading/critique. We are grateful to Alan Drury (formerly dramaturg of BBC Radio drama department) and Edward M. Cohen (former Associate Director of the Jewish Repertory Theatre in Manhattan) for dramaturgical advice; to Jean-Pierre Poirier (Paris) and Anders Lundgren (Uppsala) for offering much historical expertise; and to David Corson, Laura Linke and library staff for their generous and enthusiastic guided tour through the Cornell University Lavoisier archives.

Indispensable financial contributions by the Henry and Camille Dreyfus Foundation of New York and the Alafi Family Foundation of San Francisco made possible the fully staged workshop production run in May 2000 at the Eureka Theatre in San Francisco directed by Andrea Gordon.

A November 2000 performance of a 30-minute excerpt of the play by the American Historical Theatre (directed by Pamela Sommerfield), organized by the Chemical Heritage Foundation in Philadelphia, was generously supported by the Eugene Garfield Foundation.

The San Diego Repertory Theatre production was partially funded by the American Chemical Society, the Alafi Family Foundation, and the Henry and Camille Dreyfus Foundation, while PFIZER provided similar support toward the Royal Institution performances in London. The financial contributions by AVENTIS as well as the Fonds der Chemischen Industrie and the German Chemical Society played a crucial catalytic role in making possible the German premiere in the Würzburger Stadttheater. As playwrights, we are all too conscious of the financial and psychological value of such philanthropy and herewith express our deepest appreciation to all the above listed donors and to future ones as well.